Mathematische Optimierung und Wirtschaftsmathematik | Mathematical Optimization and Economathematics

Reihe herausgegeben von

Ralf Werner, Institut für Mathematik, Universität Augsburg, Augsburg, Bayern, Deutschland

Tobias Harks, Institut für Mathematik, Professur für Optimierung, Augsburg, Deutschland

Vladimir Shikhman, Fakultät für Mathematik, Technische Universität Chemnitz, Chemnitz, Deutschland

In der Reihe werden Arbeiten zu aktuellen Themen der mathematischen Optimierung und der Wirtschaftsmathematik publiziert. Hierbei werden sowohl Themen aus Grundlagen, Theorie und Anwendung der Wirtschafts-, Finanz- und Versicherungsmathematik als auch der Optimierung und des Operations Research behandelt. Die Publikationen sollen insbesondere neue Impulse für weitergehende Forschungsfragen liefern, oder auch zu offenen Problemstellungen aus der Anwendung Lösungsansätze anbieten. Die Reihe leistet damit einen Beitrag der Bündelung der Forschung der Optimierung und der Wirtschaftsmathematik und der sich ergebenden Perspektiven.

Weitere Bände in der Reihe http://www.springer.com/series/15822

Anja Schedel

Cost Sharing, Capacity Investment and Pricing in Networks

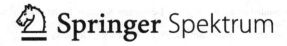 Springer Spektrum

Anja Schedel
Institute of Mathematics
University of Augsburg
Augsburg, Germany

Zugleich Dissertation, Universität Augsburg, 2020

Erstgutachter: Prof. Dr. Tobias Harks
Zweitgutachter: Prof. Dr. Martin Hoefer
Drittgutachter: Prof. Dr. Marc Uetz
Tag der mündlichen Prüfung: 17.07.2020

ISSN 2523-7926 ISSN 2523-7934 (electronic)
Mathematische Optimierung und Wirtschaftsmathematik | Mathematical Optimization and Economathematics
ISBN 978-3-658-33169-6 ISBN 978-3-658-33170-2 (eBook)
https://doi.org/10.1007/978-3-658-33170-2

Responsible Editor: Marija Kojic
This Springer Spektrum imprint is published by the registered company Springer Fachmedien Wiesbaden GmbH part of Springer Nature.
The registered company address is: Abraham-Lincoln-Str. 46, 65189 Wiesbaden, Germany

Acknowledgements

First of all, I would like to thank my supervisor, Prof. Dr. Tobias Harks, for his guidance, encouragement and patience during my dissertation project. His door was (and is) always open for me. Particularly, whenever I was stuck in my research for this thesis, he was able to provide invaluable advice. I am looking very much forward to the continued collaboration with him. Besides the scientific benefits, I also really enjoyed the barbecues and birthday parties in Munich, as well as the many coffees after lunch.

I would furthermore like to express my gratitude to the *German Research Foundation (Deutsche Forschungsgesellschaft)* for the financial support of my dissertation project.

I have greatly benefitted from the collaborations with my co-authors Manuel Surek and Prof. Dr. Martin Hoefer. In this context, I wish to particularly highlight the many discussions with Manuel Surek, which have been extremely insightful for me. Moreover, I am grateful that Prof. Dr. Martin Hoefer and Prof. Dr. Marc Uetz agreed to review this thesis. Further thank goes to Prof. Dr. Mirjam Dür for being part of my defense committee.

Next, I would like to thank all my colleagues at the chair of *Discrete Mathematics, Optimization and Operations Research* in Augsburg. Working with you has been a great pleasure for me. Special thanks go to Prof. Dr. Dirk Hachenberger and Prof. Dr. Dieter Jungnickel, whose supervisions of my bachelor and master theses laid the foundation for my dissertation project.

I owe deep gratitude to my parents and my sister. Without them, this project would not have been possible. Thank you for making me the person I am now, and for supporting me throughout my life. Last, but by no means least, I would like to thank my husband, Daniel, whose unconditional love and support has accompanied me now for over eleven years. I love you.

Augsburg   Anja Schedel
December 2019 and November 2020

Contents

Introduction

1

In this chapter, we introduce and motivate the topics of this thesis, and describe the obtained results.

1.1 Cost Sharing in Networks

The *Steiner tree problem* is a well-motivated and well-studied combinatorial optimization problem: Given an undirected graph with nonnegative edge costs, and a set of vertices of this graph, called *terminals*, find a subgraph such that all terminals are in the same connected component, and at the same time, the sum of edge costs over the contained edges is minimized. The Steiner tree problem arises in many practical applications, for example in the design of communication or transportation systems, or in physical VLSI chip design (see, for example, [73]).

A natural generalization of the Steiner tree problem is the *Steiner forest problem*: Given an undirected graph with nonnegative edge costs, and a set of pairs of vertices of this graph, called *source-sink pairs*, find a subgraph such that for each source-sink pair, there is a path connecting the corresponding two vertices, and at the same time, the sum of edge costs over the contained edges is minimized.

In this thesis, we consider a *game-theoretic variant* of the Steiner forest problem. The motivation is that networks like the Internet are built and used by a large number of agents *acting selfishly*; there is no central authority controlling the system. The analysis of strategic interaction (a *game*) between selfishly acting decision-makers (the *players* of the game) is the subject of game theory. In a so-called *network cost sharing game*, each source-sink pair is associated with a player who wants the corresponding vertices to be connected by a path. That is, each player chooses a path from the set of paths connecting her two vertices. The cost of the subgraph which is induced by the choices of the players is *shared* among the players: For each

© The Author(s), under exclusive license to Springer Fachmedien Wiesbaden GmbH, part of Springer Nature 2021
A. Schedel, *Cost Sharing, Capacity Investment and Pricing in Networks*, Mathematische Optimierung und Wirtschaftsmathematik | Mathematical Optimization and Economathematics, https://doi.org/10.1007/978-3-658-33170-2_1

edge that a player uses in her chosen path, she has to pay a share of the edge's cost. The cost shares are determined by a given *cost sharing method*, and each player chooses a path with the objective to minimize her own cost (the sum of cost shares over edges in her chosen path). A common assumption in a game-theoretic analysis of strategic interaction is that the "outcome" of a game is a *stable* situation, also called an *equilibrium*, in the sense that no player has an incentive to deviate from her decision. In this thesis, we focus on the well-known *pure Nash equilibrium*, a set of decisions, one for each player, such that a unilateral change of a player's decision is never beneficial for this player. It is well known that selfish behaviour may lead to increased total costs for pure Nash equilibria compared to a centrally enforced, *globally optimal* solution. In the context of network cost sharing games, this means that a subgraph induced by a pure Nash equilibrium may not be an optimal solution for the Steiner forest problem. Two well-known measures to quantify this efficiency loss due to selfish behaviour are the *price of anarchy* and the *price of stability*, defined by the cost ratio of a worst (most expensive) pure Nash equilibrium and a globally optimal solution, and the cost ratio between the best (least expensive) pure Nash equilibrium and a globally optimal solution, respectively ([8, 81, 94]). Note that the globally optimal solution does not depend on the cost sharing method, whereas the pure Nash equilibria may vary if we change the cost sharing method. A natural question arising in this context is: How good can an equilibrium be, if we are allowed to choose the cost sharing method (out of a reasonably defined set of methods)? We formulate this question as a two-stage problem, in which a *system designer* chooses a cost sharing method, such that the price of stability in an underlying, second-stage network cost sharing game is minimized. Equivalently, the system designer's goal is to minimize the cost of a best pure Nash equilibrium in the network cost sharing game. Implicitly, this formulation requires that the network cost sharing game induced by a chosen cost sharing method has at least one pure Nash equilibrium. Therefore, this "stability property" is contained in the axioms defining the *design space* of cost sharing methods, that is, the set of cost sharing methods that the system designer is allowed to choose. We assume here that the design space consists of all cost sharing methods satisfying the following three properties (introduced in [25]). First, the cost of each edge is exactly shared by the users of that edge (called *budget-balance*). Second, the induced network cost sharing game needs to be *stable*, that is, it has at least one pure Nash equilibrium. Finally, the cost shares on an edge remain unchanged whenever the user set of that edge is not changed. This is called *separability*. While the first two properties are natural, separability is useful for a distributed implementation of the cost sharing method in large networks (see also Subsection 3.1.3 for a more detailed discussion of these properties).

Besides for the setting described above, we also analyze the system designer's problem for some generalizations, namely we also consider directed graphs, edge costs which depend on the number of users, and edges having additional, *non-shareable* cost components (called *delays*). The motivation for considering delays is that using an edge may cause further unavoidable, and thus not shareable, costs for each user of an edge (think, for example, of costs measured by the time needed to travel along an edge in a road network). In the presence of delays, a player has to pay the cost share *plus* the complete delay of an edge, if she uses this edge in her path. A global optimum is a set of paths, one for each player, such that the sum of edge costs over the used edges, plus all delays experienced by the players, is minimized.

We now describe the main results in the context of network cost sharing games which are derived in this thesis.

Our Results

Our main results are conditions ensuring that an optimal solution of the system designer's problem yields a price of stability of 1, or, in other words, that there exists a cost sharing method such that the induced network cost sharing game has a pure Nash equilibrium which is a global optimum. We furthermore obtained some results about the efficient computation of the corresponding cost sharing methods.

Firstly, we analyze *enforceable solutions*. A *solution* consists of a path for each player, and a solution is *enforceable* if there exists a cost sharing method inducing it as a pure Nash equilibrium. We derive a characterization of enforceable solutions in terms of a linear program. The result is constructive in the sense that if a solution is enforceable, the characterization also provides a corresponding cost sharing method. Using this, the system designer's problem can be reformulated as finding an enforceable solution with smallest possible cost. We use the characterization of enforceability in all of our main results, which we describe now.

The first main result is for network cost sharing games with two players in undirected graphs. The cost structure of the edges is assumed as follows: There are no delays, and the shareable costs are fixed (i.e., the cost of an edge does not depend on the user set of that edge). For this setting, we completely characterize all graphs having the property that, for all possible edge costs, there always exists an enforceable solution which is a global optimum. Our characterization is in terms of forbidden subgraphs which we call *Bad Configurations*: The described property holds if and only if the given graph does not have a subgraph which is a Bad Configuration.

Subsequently, we study the general n-player case, and assume that the given (directed or undirected) graph has a special *series-parallel structure*. Namely, for each player, the set of paths she can choose constitutes a series-parallel graph. We show that if the shareable edge costs are nondecreasing and discrete-concave (in the number of users), and the (possibly player-specific) delays are fixed, any global optimum is enforceable. If additionally the shareable costs are fixed, we show how one can in principle compute cheap enforceable solutions in polynomial time: Starting with an arbitrary solution, we can transform this solution in polynomial time into an enforceable solution without increasing the overall cost. In this process, we also compute a corresponding cost sharing method. As a consequence, we reduced the efficient computation of a cheap enforceable solution, and a corresponding cost sharing method, to the efficient computation of a cheap solution. Unfortunately, the computation of a globally optimal solution is $\mathcal{NP}$-hard (since the well-known *uncapacitated facility location problem* is a special case). However, in the absence of delays, we show that a globally optimal, enforceable solution, together with a corresponding cost sharing method, can be computed in polynomial time.

1.2   Capacity and Price Competition in Networks

The problem described in the previous section exhibits the following two-stage structure: In the first stage, prices for using edges of an underlying network are set, and in the second stage, selfishly acting players use the edges, and pay for it. In this section, we introduce a further problem having this structure. But in contrast to the problem described before, where the prices are determined by a central authority (the system designer), we now allow for a more complex, game-theoretic setting in the first stage: We assume that edges are owned by different firms who selfishly set prices in order to maximize their own profits. A further generalization is that additionally to setting prices, firms may also invest in capacities for the owned edges, in order to increase their "attractivity" for the users of the second-stage network (think, for example in the context of a traffic network, of a wide road, which in comparison to a narrow one, is less susceptible to congestion, and consequently, more attractive). On the other hand, we impose more restrictions on the network structure, namely that the network consists of parallel edges only. Note that this restrictive assumption is also made in several other works analyzing the described or similar settings, see for example [2, 66, 75, 85, 105]. It is furthermore pointed out in [66] that there exist realistic situations which can be modeled by a network consisting of parallel edges, for example, several parallel access roads or bridges between the central district of a city and a suburb.

The motivation for the considered problem stems from privatized public roads, in which private firms build roads, and as compensation for their investment, are allowed to set prices for using the roads owned by them. As mentioned in [66], this is the key idea of the toll market in the United States, and it furthermore appears that prices are regulated by the government, in the sense that firms are not allowed to set their prices above a governmentally given price cap. As reported in [32], even road-specific price caps are common practice in the highway market of Santiago de Chile.

In the second part of this thesis, *capacity and price competition games* are used to model the above described scenario. In such a game, there is a network consisting of parallel edges between a source and a sink, and each edge corresponds to a firm owning that edge. Each firm decides about the *capacity* to be installed at her edge, and sets a *price* for the usage of that edge, which is not allowed to be larger than a given (edge-specific) price cap. Higher capacity incurs higher investment cost for a firm, but also increases the "attractivity" of the edge in the subsequent traffic routing problem. A firm can also decide to install no capacity at all, with the effect that her edge is not present in the network any more. After all firms have chosen capacities and prices for their edges, and if there is at least one edge with positive capacity, one unit of flow is to be routed from the source to the sink. Each flow particle represents an infinitesimally small customer wanting to travel from the source to the sink, thus each customer chooses one of the firms' edges. The effective cost that a customer experiences when using an edge depends on two factors: the price of the edge, and the congestion cost of the edge, where the congestion cost measures the travel time in monetary units, and is increasing in the volume of customers using the same edge, and decreasing in the installed capacity. We assume that each customer wants to selfishly minimize her own effective cost. Furthermore, the following classical assumptions in the study of traffic networks are adopted. First, a single (infinitesimally small) customer has no significant impact on the cost of an edge. Second, the outcome of the above described traffic routing problem is a *Wardrop flow*, a situation described by Wardrop's first principle [104], in which all source-sink paths carrying positive flow have equal cost, and this cost is at most the cost of any path having zero flow. In other words, all flow travels along shortest paths. Thus, a Wardrop flow can be interpreted as a pure Nash equilibrium of the traffic routing problem. Note that if the effective edge costs are strictly increasing in the amount of flow received, a Wardrop flow exists and is furthermore unique (see for example [33]). Under this assumption, the profit of a firm in a capacity and price competition game can be derived from the revenue induced by the unique Wardrop flow and the capacity investment cost, and we assume that each firm wants to selfishly maximize her own profit.

At this point, we want to draw the attention to a (crucial) assumption in the described game, namely that the demand for routing is *inelastic*. That is, as long as there exists an edge with positive capacity, the amount of customers wanting to travel through the network is fixed and does in particular not decrease with increasing effective costs. This scenario applies to realistic cases where customers are not willing to renounce travelling, like for example in higher value travel such as business or commute travel (see also Subsection 4.1.3 for a more detailed discussion). As we will see, due to this assumption, showing existence of pure Nash equilibria becomes a challenging problem in capacity and price competition games.

We now describe the main results in the context of capacity and price competition games which are derived in this thesis.

Our Results

We consider congestion cost functions which depend *linearly* on the volume of customers, and *inverse-linearly* on the installed capacity. For this setting, we show that a capacity and price competition game always has a pure Nash equilibrium. Furthermore, we show that this equilibrium is essentially unique, and its quality (in terms of utilitarian social welfare over firms and customers) can get arbitrarily bad.

The existence result requires a highly non-standard proof approach, since so-called *best responses* do not always exist. We now explain this in some more detail.

A frequently used tool to show existence of pure Nash equilibria is Kakutani's fixed point theorem [76], a generalization of the famous fixed point theorem of Brouwer to *correspondences* (set-valued mappings). Roughly speaking, Kakutani's theorem states that a correspondence defined on a compact convex set into itself has a fixed point if its values are always nonempty and convex, and the correspondence is (in some sense) continuous. This result can be used to show existence of pure Nash equilibria via the following considerations. Recall that a *strategy profile* (a set consisting of a decision or *strategy* for each player) is a pure Nash equilibrium if and only if a unilateral change of a player's strategy is never beneficial for this player. In other words, it holds for each player that her strategy is best possible under the assumption that the strategies of the other players are as given in the profile, and fixed. Such a best strategy is also called a *best response* to the given strategies of the other players. If one defines the so-called *best response correspondence* by mapping each strategy profile to all strategy profiles which can be composed of the best responses of the players to the other players' strategies in the given profile, then the pure Nash equilibria correspond to the fixed points of the best response correspondence. Thus, by showing that the best response correspondence fulfills the conditions of Kakutani's theorem, one can show the existence of pure Nash equilibria.

In the context of capacity and price competition games, this approach can not directly be applied: In the first place, we derive a complete characterization of a firm's best responses to given strategies of the other firms. In particular, we observe that best responses do not always exist (as a consequence of the assumption that the demand for routing is inelastic). Thus, the values of the best response correspondence can be empty. Since Kakutani's theorem requires nonempty values, we can not directly apply this result to show existence of pure Nash equilibria. Intuitive ideas to overcome the problem of nonexistent best responses also fail. We thus turn to a different existence result, which is due to McLennan, Monteiro and Tourky [89]. They show existence of pure Nash equilibria for games being *C-secure* at strategy profiles which are no pure Nash equilibria. The used concept of C-security is related to a "robustness" of strategies, combined with a "separation" of current strategies from "better" strategies (see Chapter 4 for more intuition, and all formal definitions). In this thesis, we show that the considered capacity and price competition games fulfill the conditions of McLennan et al.'s result, which leads to the existence of pure Nash equilibria. The corresponding proof, and also the proofs of the uniqueness and quality results, rely heavily on the characterization of best responses which we derived in the first place.

1.3 Thesis Organization

The organization of this thesis is as follows. In the next chapter (Chapter 2), we review some notations about graphs and games, as well as some results about correspondences which are used in the subsequent chapters. Chapter 3 is devoted to the analysis of cost sharing in networks, whereas Chapter 4 deals with capacity and price competition games. Finally, Chapter 5 contains some concluding remarks and open problems.

Preliminaries

In this chapter, we briefly review some notations according to graphs and games which are needed in this thesis, since in particular graph theoretic notations tend to vary in the literature. Furthermore, we review some results with respect to correspondences which are useful in the analysis of capacity and price competition games.

2.1 Graphs

In this thesis, a graph $G = (V, E)$ has a *finite*, nonempty set V of vertices (or nodes), and a *finite* set of edges E, and E does not contain loops, whereas parallel edges are allowed. Furthermore, a graph can be a directed or an undirected graph. We write an undirected edge connecting the two vertices u and v by $u - v$, and the directed version from u to v by $u \rightarrow v$. In both cases, we call u and v the *endnodes* of the edge; and in the directed case, u is additionally denoted as the *startnode* of the edge.

In an undirected graph $G = (V, E)$, a *walk* W is a sequence $v_1 \overset{e_1}{-} v_2 \overset{e_2}{-} \ldots \overset{e_{k-1}}{-} v_k$ (with $k \geq 1$), where $v_1, v_2, \ldots, v_k$ are vertices of G, and e_i is an edge in G with endnodes v_i, v_{i+1} for any $i = 1, \ldots, k - 1$. In a directed graph $G = (V, E)$, a *walk* W is analogously defined as a sequence $v_1 \overset{e_1}{\rightarrow} v_2 \overset{e_2}{\rightarrow} \ldots \overset{e_{k-1}}{\rightarrow} v_k$ (with $k \geq 1$), where $v_1, v_2, \ldots, v_k$ are vertices of G, and e_i is an edge in G with startnode v_i and endnode v_{i+1} for any $i = 1, \ldots, k - 1$. For a (directed or undirected) walk W, the nodes v_1 and v_k are called the *endnodes of* W; and if W is directed, v_1 is additionally specified as the *startnode of* W. We say that a node v (an edge e) is contained or used in the walk W if $v = v_i$ for some $i \in \{1, \ldots, k\}$ ($e = e_i$ for some $i \in \{1, \ldots, k - 1\}$), and write $v \in W$ ($e \in W$). Denote the set of used nodes (edges) in W with $V(W)$ ($E(W)$). If $v_1 = v_k$ holds in a walk W, we say that W is a *closed*

© The Author(s), under exclusive license to Springer Fachmedien Wiesbaden GmbH, part of Springer Nature 2021
A. Schedel, *Cost Sharing, Capacity Investment and Pricing in Networks*, Mathematische Optimierung und Wirtschaftsmathematik I Mathematical Optimization and Economathematics, https://doi.org/10.1007/978-3-658-33170-2_2

walk. A walk W in G such that the nodes v_i, $i = 1, \ldots, k$, are distinct, is called a *path* (in the literature, this is sometimes called a simple path). A closed walk W in G with $k \geq 3$ such that the nodes v_i, $i = 1, \ldots, k-1$, are distinct, and also the edges e_i, $i = 1, \ldots, k-1$, are distinct, is called a *cycle* (sometimes denoted as a simple cycle in the literature). A *subpath* of a given path P (or cycle C) is a path P' with $E(P') \subseteq E(P)$ ($E(P') \subsetneq E(C)$, respectively). Two paths P and Q in G are *node-disjoint* (*edge-disjoint*) if the two sets of used nodes (edges) $V(P)$ and $V(Q)$ ($E(P)$ and $E(Q)$) are disjoint. Furthermore, P and Q are *internal node-disjoint*, if the two sets of nodes $V(P) \setminus \{v_1, v_k\}$ and $V(Q) \setminus \{u_1, u_\ell\}$ are disjoint, where v_1 and v_k are the two endnodes of W, and u_1 and u_ℓ are the two endnodes of Q.

Given a subset of nodes $V' \subseteq V$ of a graph $G = (V, E)$, denote with $E(V') := \{e \in E : \text{both endnodes of } e \text{ are contained in } V'\}$ the subset of edges between vertices in V'. A graph $G' = (V', E')$ with $V' \subseteq V$ and $E' \subseteq E(V')$ is called a *subgraph* of G. In particular, the graph $(V', E(V'))$ is called the *subgraph induced by V'*. On the other hand, given a subset of edges $E' \subseteq E$, define $V(E') := \{v \in V : v \text{ is endnode of some edge in } E'\}$ as the set of endnodes of edges in E', and denote the graph $(V(E'), E')$ as the *subgraph induced by E'*. Finally, given a set of paths $\mathcal{Q}$ in G, the *subgraph induced by $\mathcal{Q}$* is defined as the subgraph induced by the set of edges which are used in $\mathcal{Q}$, that is, $\{e \in E : e \text{ is used in some path } P \in \mathcal{Q}\}$.

2.2   Games

A (strategic) *game* is given by a finite set N of *players* with $|N| \geq 2$, a nonempty set S_i of *strategies* for each player $i \in N$, and a *(cost or utility) function* $f_i : S \to \mathbb{R}$ for each player $i \in N$, where $S := \times_{i \in N} S_i$. In a utility maximization game, each player $i \in N$ seeks to maximize her utility function f_i, whereas in a cost minimization game, she wants to minimize f_i.

An element $s \in S$ consists of a strategy s_i for each player $i \in N$ and is called a *strategy profile*. Using standard notation in game theory, for a strategy profile $s \in S$ we denote by $(s_i', s_{-i}) \in S$ the strategy profile that arises from s if player i deviates to strategy $s_i' \in S_i$, whereas all other players $j \in N \setminus \{i\}$ keep their strategies in s. That is, s_{-i} consists of the strategies s_j for all players $j \in N \setminus \{i\}$ different from i.

A commonly used solution concept in non-cooperative game theory are *pure Nash equilibria (PNE)*. A PNE is a strategy profile $s \in S$ such that no player has an incentive to unilaterally deviate, since for any player $i \in N$, her current strategy s_i is the best she can do provided the other players do not change their strategies. Formally, a strategy profile $s \in S$ is a PNE in a utility maximization game if

$$f_i(s_i', s_{-i}) \leq f_i(s_i, s_{-i}) \text{ for all } s_i' \in \mathcal{S}_i$$

holds for each player $i \in N$, and s is a PNE in a cost minimization game if

$$f_i(s_i', s_{-i}) \geq f_i(s_i, s_{-i}) \text{ for all } s_i' \in \mathcal{S}_i$$

holds for each player $i \in N$. A different way to characterize PNE is via *best response correspondences* (see Section 2.3 for some basic notations about correspondences). For a player $i \in N$, define the best response correspondence of player i, for a utility maximization game, by

$$\text{BR}_i(s_{-i}) := \{s_i \in \mathcal{S}_i : s_i \text{ maximizes } f_i(\cdot, s_{-i}) \text{ over } \mathcal{S}_i\},$$

and for a cost minimization game, by

$$\text{BR}_i(s_{-i}) := \{s_i \in \mathcal{S}_i : s_i \text{ minimizes } f_i(\cdot, s_{-i}) \text{ over } \mathcal{S}_i\},$$

where s_{-i} are given strategies of the other players. A strategy $s_i \in \text{BR}_i(s_{-i})$ is called a *best response* of player i to s_{-i}. The (global) best response correspondence is defined by

$$\text{BR}(s) := \{s' \in \mathcal{S} : s_i' \in \text{BR}_i(s_{-i}) \text{ for each } i \in N\},$$

where $s \in \mathcal{S}$. Obviously, a strategy profile s is a PNE if and only if $s_i \in \text{BR}_i(s_{-i})$ holds for each player $i \in N$, and this is equivalent to the fact that s is a fixed point of the best response correspondence BR.

Given a cost (or utility) function over the set of strategy profiles, one can measure the quality loss which arises due to selfish behaviour (examples for frequently used cost functions are the sum of all players' costs, or the maximum cost of a player). Two well-known concepts are the *price of anarchy (PoA)* and the *price of stability (PoS)*, introduced by Koutsoupias and Papadimitriou [81, 94] and Anshelevich et al. [8], respectively. Roughly, the PoA is the ratio between the cost of a *worst* PNE and the cost of a optimal strategy profile (that would be achieved if the players could be forced to choose a specific strategy), and the PoS is the ratio between the cost of a *best* PNE and the cost of a socially optimal strategy profile (where "worst", "best" and "optimal" is with respect to the cost function over the set of strategy profiles). Formally, given a game G having at least one PNE, and a cost function $C : \mathcal{S} \to \mathbb{R}$ over the set of strategy profiles which one wants to minimize, the PoA and PoS of the game G are defined as

$$\text{PoA}(G) := \frac{\sup\{C(s) : s \text{ is PNE of the game } G\}}{\inf\{C(s) : s \in \mathcal{S}\}}$$

and

$$\text{PoS}(G) := \frac{\inf\{C(s) : s \text{ is PNE of the game } G\}}{\inf\{C(s) : s \in \mathcal{S}\}}$$

respectively. In the literature, PoA and PoS of games having no equilibria are sometimes defined as ∞. Besides the PoA or PoS of a specific game, one is often interested in the *worst-case PoA*, or the *worst-case PoS*, of an entire *class of games*. That is, given a class of games $\mathcal{G}$, one is interested in

$$\text{PoA}(\mathcal{G}) := \sup\{\text{PoA}(G) : G \in \mathcal{G}\} \text{ and } \text{PoS}(\mathcal{G}) := \sup\{\text{PoS}(G) : G \in \mathcal{G}\}.$$

2.3   Correspondences

In this section, we briefly review some notations and results according to correspondences which are used in Chapter 4 of this thesis. The notations and the formulations of the theorems follow, basically, the books [5, 21], but are only stated in the context of Euclidean vector spaces.

A *correspondence* γ from a set $X \subseteq \mathbb{R}^m$ into a set $Y \subseteq \mathbb{R}^k$ maps each element $x \in X$ to a subset $\gamma(x)$ of Y. We write $\gamma : X \twoheadrightarrow Y$. A correspondence γ is *closed at* $x \in X$, if whenever $x^n \to x$, $y^n \to y$ with $x^n \in X$ and $y^n \in \gamma(x^n)$ for all n, then $y \in \gamma(x)$. If γ is closed at every $x \in X$, we say that γ is *closed*. A *fixed point* of a correspondence γ is an element $x \in X$ with $x \in \gamma(x)$.

An important theorem, which is widely used in game theory to prove existence of PNE (via the existence of a fixed point of the best response correspondence), is *Kakutani's fixed point theorem* ([76], see also the generalizations due to Fan [40] and Glicksberg [52]). It can be stated as follows:

Theorem 2.3.1 (Kakutani's fixed point theorem [76]). *Let $\emptyset \neq X \subseteq \mathbb{R}^m$ be compact and convex, and $\gamma : X \twoheadrightarrow X$ a closed correspondence with $\gamma(x)$ nonempty and convex for each $x \in X$. Then γ has a fixed point.*

A further useful theorem is the so-called *theorem of the maximum* due to Berge [14]. Roughly speaking, it states that the set of optimal solutions of a maximization problem, as well as the optimal objective function value, behaves continuously if the feasible set varies in a continuous way. To state the theorem formally, we need to introduce further notions related to continuity of correspondences. The formulation of the theorem, as well as the following definitions, are taken from [5]. A *neighborhood of a set* A is a set B for which there is an open set V satisfying $A \subseteq V \subseteq B$. An open set V that satisfies $A \subseteq V$ is called an *open neighborhood of* A. A correspondence γ is *upper hemi-continuous at* $x \in X$ if for every open neighborhood V of $\gamma(x)$, the set $\{x' \in X : \gamma(x') \subseteq V\}$ is a neighborhood of $\{x\}$, and γ is *lower hemi-continuous at* $x \in X$ if for every open set U with $\gamma(x) \cap U \neq \emptyset$, the set $\{x' \in X : \gamma(x') \cap U \neq \emptyset\}$ is a neighborhood of $\{x\}$. If γ is both upper and lower hemi-continuous at $x \in X$, we say that γ is *continuous at* x. If γ is upper hemicontinuous (lower hemicontinuous, continuous) at every $x \in X$, we say that γ is *upper hemicontinuous* (*lower hemicontinuous, continuous*). Note that in the literature, one sometimes speaks of *semi-* instead of hemi-continuity. Furthermore, in case that γ is a single-valued correspondence, the continuity of the corresponding function is equivalent to the upper (lower) hemi-continuity of γ. We can now state the theorem of the maximum as follows.

Theorem 2.3.2 (Berge's theorem of the maximum [14]). *Let* $\emptyset \neq X \subseteq \mathbb{R}^m$, $\emptyset \neq Y \subseteq \mathbb{R}^k$, *and* $\gamma : X \longrightarrow\longrightarrow Y$ *a continuous correspondence with* $\gamma(x)$ *nonempty and compact for each* $x \in X$. *Furthermore,* $f : X \times Y \to \mathbb{R}$ *is a continuous function. Define* $m : X \to \mathbb{R}$ *by*

$$m(x) := \max\{f(x, y) : y \in \gamma(x)\},$$

and $\mu : X \longrightarrow\longrightarrow Y$ *by*

$$\mu(x) := \{y \in \gamma(x) : f(x, y) = m(x)\}.$$

Then the "value function" m *is continuous, and the "argmax correspondence"* μ *is upper hemi-continuous with* $\mu(x)$ *nonempty and compact for each* $x \in X$.

Cost Sharing in Networks 3

3.1 Introduction

In the following chapter of this thesis, we consider a generalization of the well-known combinatorial Steiner forest problem in a game-theoretic context. In the Steiner forest problem, we are given an undirected graph with nonnegative edge costs, and a set of pairs of vertices called source-sink pairs. The aim is to find a cost-minimal subgraph with the property that there is a path between the source and the sink for each source-sink pair. Motivated by the fact that many networks (like the Internet) are build and used by *selfishly* acting agents, we consider a game-theoretic variant of the Steiner forest problem. We assume that each source-sink pair corresponds to a player who chooses a path connecting the source with the sink. After all players have chosen a path, the cost of the resulting subgraph is shared among the players: For each edge that a player uses in her path, she has to pay a cost share which is determined by a (given) *cost sharing method*. Assuming that each player chooses a path with the objective to minimize the sum of cost shares that she has to pay, this defines a strategic game, which we call a *network cost sharing game*. In fact, we consider a generalization of the above described setting, since we also allow for directed graphs, and edge costs which depend on the number of players using an edge. Furthermore, edges additionally have *non-shareable* cost components, called *delays*, which have to be paid by each player using the edge. This is for example motivated by costs resulting from the time which is needed to travel through an edge in a road network. In the following chapter of this thesis, we consider the described network cost sharing games from the perspective of a system designer, e.g. a governmental regulator, who is allowed to *choose* the cost sharing method for a network cost sharing game. The objective of the system designer is to minimize the overall cost of the best PNE in the game induced by the

© The Author(s), under exclusive license to Springer Fachmedien Wiesbaden GmbH, part of Springer Nature 2021
A. Schedel, *Cost Sharing, Capacity Investment and Pricing in Networks*, Mathematische Optimierung und Wirtschaftsmathematik | Mathematical Optimization and Economathematics, https://doi.org/10.1007/978-3-658-33170-2_3

chosen cost sharing method. Obviously, this problem is greatly influenced by the set of cost sharing methods from which the system designer is allowed to choose. Here, we allow for all cost sharing methods which are budget-balanced, stable and separable (properties introduced in a seminal paper of Chen et al. [25]). Budget-balance requires that the cost of an edge is exactly covered by the collected cost shares. Stability means that the induced network cost sharing game has at least one PNE. Finally, separability requires that the cost shares on an edge remain unchanged whenever the user set of this edge is not changed (see Definition 3.1.1 for a formal definition, and Subsection 3.1.3 for a discussion of these properties). Altogether, the problem considered in this chapter is the one of a system designer who wants to minimize the cost of a best PNE in a network cost sharing game, and to this end, is allowed to choose a budget-balanced, stable and separable cost sharing method. Our main results are conditions ensuring that an optimal solution of the system designer's problem induces an *optimal* strategy profile as a PNE (where an optimal profile is minimal with respect to the total cost). We furthermore obtained some results about the efficient computation of the corresponding cost sharing methods.

The organization of the chapter is as follows. In the remaining part of Section 3.1, we formally describe the problem, followed by a summary of our main results and the used proof techniques, and close with a brief discussion of the model and related work. The sections 3.2, 3.3 and 3.4 then contain the technical presentation of our results. In Section 3.2 we present a simple, yet useful tool which is the base of our main results. The main results are then contained in Section 3.3 (for the special case of two players) and Section 3.4 (for the general case with $n \geq 2$ players). Finally, Section 3.5 contains some bibliographic notes.

3.1.1   Problem Description

In a *network cost sharing game*, we are given a (directed or undirected) graph $G = (V, E)$ with vertex set V and edge set E. The graph G does not contain loops, but parallel edges are allowed. Furthermore, there are $n \geq 2$ pairs $(s_1, t_1), \ldots, (s_n, t_n)$ of vertices in G, called source-sink pairs. Each such pair (s_i, t_i) corresponds to a player $i \in N := \{1, \ldots, n\}$ whose strategy set is given by the set $\mathcal{P}_i$ of (s_i, t_i)-*paths* in G. Here, an (s_i, t_i)-path in G denotes a path in G with endvertices s_i and t_i if G is undirected, and a path in G with startvertex s_i and endvertex t_i, if G is directed. Without loss of generality, assume that $s_i \neq t_i$ and $\mathcal{P}_i \neq \emptyset$ for each player $i \in N$.[1] A

---

[1] If there is a player $i \in N$ with $s_i = t_i$, or with no (s_i, t_i)-path (this can easily be verified by breadth first search), we can omit this player from the game.

strategy profile $P = (P_1, \ldots, P_n)$ consists of an (s_i, t_i)-path $P_i \in \mathcal{P}_i$ for each player $i \in N$. Denote the set of strategy profiles by $\mathcal{P}$. For each edge $e \in E$ and each strategy profile $P \in \mathcal{P}$, denote the set of players using e in P by $N_e(P) := \{i \in N : e \in P_i\}$, and denote by $n_e(P) := |N_e(P)|$ the corresponding number of users. The *shareable cost* of edge $e \in E$ under $P \in \mathcal{P}$ depends on the number of users $n_e(P)$ and is given by a nonnegative function $c_e : \mathbb{N} := \{0, 1, 2, 3, \ldots\} \to \mathbb{R}_{\geq 0} := \{r \in \mathbb{R} : r \geq 0\}$. To simplify notation, we write $c_e(P) := c_e(n_e(P))$, and $c := (c_e)_{e \in E}$ for the vector of all cost functions. How the cost $c_e(P)$ of edge $e \in E$ is shared is given by a *cost sharing method* ξ_e which assigns, for each strategy profile P with $N_e(P) \neq \emptyset$, nonnegative *cost shares* $\xi_{i,e}(P) \in \mathbb{R}_{\geq 0}$ for each $i \in N_e(P)$, i.e., each user of e. We write $\xi := (\xi_e)_{e \in E}$ for the vector of all cost sharing methods. Additionally to the shareable cost, each edge $e \in E$ also has a nonnegative *player-specific delay* $d_{i,e} \in \mathbb{R}_{\geq 0}$ for each player $i \in N$.[2] This delay is not shareable, i.e., if player i uses edge e in her path, then she has to pay $d_{i,e}$ completely. We write $d := (d_{i,e})_{i \in N, e \in E}$ for the vector consisting of all delays. Altogether, the cost that player $i \in N$ has to pay under strategy profile $P \in \mathcal{P}$ is given by

$$\xi_i(P) := \sum_{e \in P_i} \left(\xi_{i,e}(P) + d_{i,e} \right).$$

Assuming that each player $i \in N$ wants to minimize her own cost, we have thus defined a strategic (cost minimization) game $\mathcal{G}$ which we call a *network cost sharing game*, and we write $\mathcal{G} = (G, (s_i, t_i)_{i \in N}, c, d, \xi)$. We furthermore call $\mathcal{N} := (G, (s_i, t_i)_{i \in N}, c, d)$ the *network of the game* $\mathcal{G}$, and ξ the *cost sharing method of* $\mathcal{G}$, and we also write $\mathcal{G}$ in the form $(\mathcal{N}, \xi)$.

The *underlying optimization problem* of a network cost sharing game $\mathcal{G}$ is the problem of computing a cost-minimal strategy profile, where the cost of a strategy profile $P \in \mathcal{P}$ is defined by the sum of shareable edge costs over all used edges, plus all player-specific delays that the players experience, i.e.,

$$C(P) := \sum_{e \in E : N_e(P) \neq \emptyset} c_e(n_e(P)) + \sum_{i \in N} \sum_{e \in P_i} d_{i,e}.$$

Note that the cost of a strategy profile does not depend on the cost sharing method. An *optimal strategy profile* is a cost-minimal strategy profile.

---

[2]Note that we assume, in contrast to the shareable costs, that the delays are *fixed*, i.e., do not depend on the number of players using edge e (or, more general, on P).

In order to be practically relevant, cost sharing methods need to satisfy several desiderata. Chen et al. [25] axiomatized cost sharing methods by the following three fundamental properties (see also [30, 103]; for a discussion of these properties, see Subsection 3.1.3).

Definition 3.1.1 (properties of cost sharing methods). Given a network cost sharing game $\mathcal{G} = (G, (s_i, t_i)_{i \in N}, c, d, \xi) = (\mathcal{N}, \xi)$, we call the cost sharing method ξ

1. *budget-balanced*, if for each strategy profile $P \in \mathcal{P}$ and each edge $e \in E$ which is used in P, i.e., with $N_e(P) \neq \emptyset$, the cost of e is completely paid by the users of e, i.e.,
$$\sum_{i \in N_e(P)} \xi_{i,e}(P) = c_e(P).$$

2. *stable*, if $\mathcal{G}$ has at least one PNE.
3. *separable*, if for each edge $e \in E$, the cost shares on edge e only depend on the user set of e, i.e., $N_e(P) = N_e(P')$ implies $\xi_{i,e}(P) = \xi_{i,e}(P')$ for each $i \in N_e(P)$ and $P, P' \in \mathcal{P}$.

Given a stable cost sharing method ξ, the *PoS of* ξ is defined as the PoS of the game $\mathcal{G} = (\mathcal{N}, \xi)$ induced by ξ (recall that the PoS of $\mathcal{G}$ is the cost of a best (cost-minimal) PNE of $\mathcal{G}$ divided by the cost of an optimal strategy profile in $\mathcal{G}$).

Perhaps the most prominent cost sharing method is *fair sharing*, also called *Shapley cost sharing*[3]. Here, the cost of each used edge is allocated in equal shares to the players using it, i.e., for each strategy profile $P \in \mathcal{P}$ and edge $e \in E$ with $N_e(P) \neq \emptyset$, it assigns $\xi_{i,e}(P) = c_e(P)/n_e(P)$ for all $i \in N_e(P)$. Obviously, Shapley cost sharing is budget-balanced and separable. Furthermore, for the case of player-independent delays, i.e., $d_{i,e} = d_e$ for all $i \in N, e \in E$, Shapley cost sharing is stable (since the induced game is a congestion game, see [9, 91, 97]). To illustrate the introduced game, we now consider an example with Shapley cost sharing as cost sharing method.

Example 3.1.2 (Shapley cost sharing). We consider a simple network cost sharing game for two players, i.e., $N = \{1, 2\}$, and we assume that the shareable edge costs are fixed, i.e., constant functions. Additionally, let all player-specific delays

---

[3]Note that assigning $c_e(P)/n_e(P)$ to each player in $N_e(P)$ is the same as the Shapley value of the cooperative game with players $N_e(P)$ and characteristic function v defined by $v(N') := c_e(|N'|)$ for all $\emptyset \neq N' \subseteq N_e(P)$ (and $v(\emptyset) := 0$).

be zero. Figure 3.1a shows the graph G with vertex set $V = \{s_1, t_1, t_2\}$, edge set $E = \{e, f, g\}$ and the source-sink pairs (s_1, t_1), (s_2, t_2). The edges are labelled by their name, followed by their fixed cost, where $0 < \varepsilon < 1/3$.

The strategies of Player 1 are $P_1^1 = s_1 - t_1$ and $P_1^2 = s_1 - t_2 - t_1$. For Player 2, there are $P_2^1 = s_2 - t_2$ and $P_2^2 = s_2 - t_1 - t_2$. Thus, we have four different strategy profiles (see Figure 3.1), namely

$$P^1 = (P_1^1, P_2^1) \text{ with cost } C(P^1) = c_e + c_f = 4 + \varepsilon,$$
$$P^2 = (P_1^1, P_2^2) \text{ with cost } C(P^2) = c_e + c_g = 3 + 3\varepsilon,$$
$$P^3 = (P_1^2, P_2^1) \text{ with cost } C(P^3) = c_f + c_g = 3 + 2\varepsilon \text{ and}$$
$$P^4 = (P_1^2, P_2^2) \text{ with cost } C(P^4) = c_e + c_f + c_g = 5 + 3\varepsilon.$$

(recall that the cost of a strategy profile is given by the sum of edge costs over all used edges). Note that P^3 is the unique optimal strategy profile.

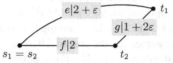

(a) Network of the game (edges are labelled by their name, followed by their fixed cost).

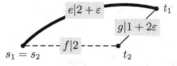

(b) The profile P^1: P_1^1 thick, P_2^1 dashed.

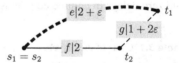

(c) The profile P^2: P_1^1 thick, P_2^2 dashed.

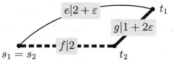

(d) The profile P^3: P_2^2 thick, P_2^1 dashed.

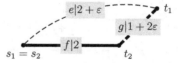

(e) The profile P^4: P_1^2 thick, P_2^2 dashed.

Fig. 3.1 A network cost sharing game with Shapley cost sharing

We now consider the game induced by Shapley cost sharing. Table 3.1 shows the costs of the players for all possible strategy profiles. Recall that for Shapley cost sharing, whenever a player uses an edge alone, she has to pay the edge completely, and if both players use an edge, each player has to pay half of the cost. For example, the cost that Player 1 has to pay under P^2 is $\xi_1(P^2) = \xi_{1,e}(P^2) = c_e/2 = 1 + \varepsilon/2$ since both players use edge e. For Player 2, we get $\xi_2(P^2) = \xi_{2,e}(P^2) + \xi_{2,g}(P^2) = c_e/2 + c_g = 1 + \varepsilon/2 + 1 + 2\varepsilon = 2 + 5\varepsilon/2$ since both players use e and Player 2 uses g alone.

Next, we analyze which of the strategy profiles are PNE. Recall that a strategy profile is a PNE if and only if no player can strictly decrease her cost by a unilateral deviation. Consider P^1. If Player 1 deviates to (her only alternative) P_1^2, the resulting strategy profile is P^3, where she has to pay $2 + 2\varepsilon$ which is more than $2 + \varepsilon$. Thus, there is no improving deviation for Player 1. If Player 2 changes her strategy to P_2^2, which yields the strategy profile P^2, she also does not decrease her cost, since $2 + 5\varepsilon/2 > 2$. Therefore, Player 2 also cannot decrease her cost by a unilateral deviation, which overall shows that P^1 is a PNE. In fact, P^1 is the only PNE of the game (we provide improving deviations for the other strategy profiles in Table 3.1). In particular, the unique optimal strategy profile P^3 is no PNE, since Player 1 can decrease her cost from $2 + 2\varepsilon$ to $2 + \varepsilon$ by changing her strategy from P_1^2 to P_1^1. This shows that the PoS of the game, i.e., the cost of a best (cost-minimal) PNE, divided by the cost of an optimal solution, is $\frac{4+\varepsilon}{3+2\varepsilon}$. For ε going to zero, this ratio tends to $4/3$, which is actually the worst-case over all network cost sharing games induced by Shapley cost sharing with two players, constant nonnegative edge cost functions and zero delays in undirected graphs (see [9, 27]).

Table 3.1 Costs of the players

P	$\xi_1(P)$	$\xi_2(P)$	PNE?
P^1	$2 + \varepsilon$	2	Yes
P^2	$1 + \frac{\varepsilon}{2}$	$2 + \frac{5\varepsilon}{2}$	No ($P_2^2 \to P_2^1$)
P^3	$2 + 2\varepsilon$	1	No ($P_1^2 \to P_1^1$)
P^4	$\frac{5}{2} + \varepsilon$	$\frac{5}{2} + 2\varepsilon$	No ($P_1^2 \to P_1^1$ or $P_2^2 \to P_2^1$)

Shapley cost sharing is simple and has been studied intensively, but there are several significant drawbacks (see Subsection 3.1.3). One disadvantage, which we have already seen in Example 3.1.2, is that the PoS of an induced network cost sharing game is not necessarily 1. Since the optimality of a strategy profile does not depend on the cost sharing method, the following question naturally arises: Is there a (different) budget-balanced, stable and separable cost sharing method, such that the PoS is 1? This question can be embedded in the problem of a *system designer*, who is allowed to choose the (budget-balanced, stable and separable) cost sharing method, with the objective to minimize the PoS in the resulting network cost sharing game. The question above then reads: Does an optimal solution of the system designer's problem lead to a PoS of 1? At least for the network presented in Example 3.1.2, this holds true:

Example 3.1.2 (continued). We want to design a budget-balanced and separable cost sharing method ξ such that P^3 is a PNE of the induced game. Note that budget-balance implies that whenever a player uses an edge alone, she has to pay this edge completely, and if both players use an edge, it suffices to determine the cost share of one player, since the cost share of the other player is then uniquely determined by the difference between the edge cost and the cost share of the first player. Finally, recall that separability means that the cost shares on an edge only depend on the user set of this edge. Therefore, to determine ξ completely, it suffices to define, for each edge h, the cost share of Player 1 if both players use edge h. Denote this value by ξ_h, with $\xi_h \in [0, c_h]$. We have to make sure that $P^3 = (P_1^2, P_2^1)$ is a PNE. For the costs of the players under P^3, we get

$$\xi_1(P^3) = \xi_{1,f}(P^3) + \xi_{1,g}(P^3) = \xi_f + 1 + 2\varepsilon,$$
$$\xi_2(P^3) = \xi_{2,f}(P^3) = c_f - \xi_f = 2 - \xi_f \leq 2.$$

If Player 1 deviates to P_1^1, she has to pay $c_e = 2 + \varepsilon$. Since this should not be an improving deviation, we get $\xi_f + 1 + 2\varepsilon \leq 2 + \varepsilon$, i.e., $\xi_f \leq 1 - \varepsilon$. If Player 2 deviates to P_2^2, she has to pay $c_e + c_g - \xi_g \geq c_e = 2 + \varepsilon > 2 \geq \xi_2(P^3)$. Thus, independently of the values ξ_h, Player 2 does not have an improving deviation. Altogether, any choice of $\xi_e \in [0, 2 + \varepsilon]$, $\xi_f \in [0, 1 - \varepsilon]$ and $\xi_g \in [0, 1 + 2\varepsilon]$ yields a cost sharing method having P^3 as a PNE, and thus with a PoS of 1. For completeness, Figure 3.2b shows one example of such a cost sharing method, where the edges h are labelled by their cost c_h, followed by ξ_h.

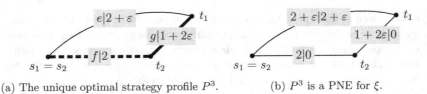

(a) The unique optimal strategy profile P^3.   (b) P^3 is a PNE for ξ.

Fig. 3.2 A cost sharing method having P^3 as a PNE

In the above example, we were able to show that a budget-balanced and separable cost sharing method inducing an optimal strategy profile as a PNE exists. In other words, an optimal solution of the system designer's problem led to a network cost sharing game with PoS equal to 1. Naturally, one can ask if a PoS of 1 can *always* be achieved. In general, the answer to this question is "no": Not even for the special case with only two players, constant shareable edge cost functions and zero delays (see Section 3.3), one can guarantee the existence of a budget-balanced and separable cost sharing method inducing an optimal strategy profile as a PNE. From the viewpoint of the system designer who is allowed to choose the cost sharing method in order to minimize the PoS, this means that an optimal cost sharing method does not necessarily lead to a PoS of 1.

In this chapter, we focus on finding conditions on the network which *guarantee* the existence of a budget-balanced and separable cost sharing method with PoS of 1. In particular, we want to find such conditions which do not depend on the costs and delays (for example, such conditions are useful in scenarios where costs or delays are not known, or only known probabilistically). To this end, it is useful to introduce the notions of *enforceable* strategy profiles, i.e., profiles which can actually appear as PNE, and of *efficient* graphs, ensuring the existence of an enforceable, optimal strategy profile independently of (certain sets of) costs and delays:

Definition 3.1.3 (enforceable strategy profile). Let $\mathcal{N}$ be a network. A strategy profile $P \in \mathcal{P}$ is called *enforceable*, if there is a budget-balanced and separable cost sharing method ξ such that P is a PNE of the game $(\mathcal{N}, \xi)$. If $P \in \mathcal{P}$ is enforceable and ξ is a budget-balanced and separable cost sharing method such that P is a PNE of the game $(\mathcal{N}, \xi)$, we say that ξ *enforces* P.[4]

---

[4]Note that in general, a cost sharing method enforcing P is not unique: In Example 3.1.2, we showed that there are (infinitely) many cost sharing methods enforcing P^3.

Definition 3.1.4 ((strongly) efficient graph). Let $G = (V, E)$ be a graph with source-sink pairs $(s_i, t_i)_{i \in N}$. Furthermore, let C be a set consisting of edge cost functions $c = (c_e)_{e \in E}$, and $\mathcal{D}$ be a set consisting of delays $d = (d_{i,e})_{i \in N, e \in E}$.

We call $(G, (s_i, t_i)_{i \in N})$ *efficient for* $(C, \mathcal{D})$, if, for every $c \in C$ and $d \in \mathcal{D}$, there is an optimal strategy profile of the network $\mathcal{N} = (G, (s_i, t_i)_{i \in N}, c, d)$ which is enforceable.

Furthermore, we call $(G, (s_i, t_i)_{i \in N})$ *strongly efficient for* $(C, \mathcal{D})$, if, for every $c \in C$ and $d \in \mathcal{D}$, *every* optimal strategy profile of the network $\mathcal{N} = (G, (s_i, t_i)_{i \in N}, c, d)$ is enforceable.

In this chapter of the thesis, we focus on the analysis of efficient graphs. Given an efficient graph (for some sets of edge costs and delays), an optimal solution of the system designer's problem always leads to a PoS of 1. In particular, we are interested in the efficient computation of optimal (or at least good) solutions for the problem of the system designer.

3.1.2 Main Results and Proof Techniques

We now describe the results with respect to the introduced cost sharing problems which are derived in this thesis.

In Section 3.2, we present a characterization of enforceable strategy profiles, which was already used in [10, 67, 103] for slightly different settings. The characterization is in terms of a linear program (LP). A given strategy profile is enforceable if and only if the LP has an optimal solution which satisfies a certain property. In particular, if the profile is enforceable, the LP provides a corresponding cost sharing method. This way, the system designer's problem can be reformulated as finding an enforceable strategy profile with smallest possible cost. The described LP-characterization of enforceability is the basis for obtaining all of our main results, which we describe now.

Our first main result, presented in Section 3.3, is for two-player games in undirected graphs with constant edge cost functions and zero delays. For this setting, we completely characterize efficient graphs. Our characterization is in terms of forbidden subgraphs, so-called *Bad Configurations*: The undirected graph G with source-sink pairs (s_1, t_1), (s_2, t_2) of the two players is efficient for constant edge costs and zero delays if and only if there is no subgraph which is a Bad Configuration. As a byproduct of the proof of this result, we also get that efficiency is equivalent to strong efficiency (for the considered special case). Whereas the proof of the "only-if" part of the characterization is quite straightforward, the proof of the other direction (i.e., a graph is efficient if it does not contain a Bad Configuration as a subgraph) is much more complicated. The general proof approach is by contradiction, and makes use of the LP-characterization of enforceability from Section 3.2, but besides this, several additional ideas are needed, and a lot of different cases have to be analyzed.

In Section 3.4, we study the general case of n players. We show that if $(G, (s_i, t_i)_{i \in N})$ is n-series-parallel, meaning that for each player $i \in N$, the set of (s_i, t_i)-paths forms a series-parallel graph, then $(G, (s_i, t_i)_{i \in N})$ is strongly efficient for *nondecreasing and discrete-concave*[5] edge costs and arbitrary delays. Our proof of this result is constructive: We present Algorithm n-SEPA which takes as input an arbitrary strategy profile P and outputs P, if P is enforceable, and otherwise computes an enforceable strategy profile with strictly smaller cost than P. In particular, this shows the above stated result: If there exists an optimal, but not enforceable strategy profile P, we can use P as input for Algorithm n-SEPA, which then computes a strategy profile with strictly smaller cost than P. Since this is a contradiction to P's optimality, we conclude that any optimal strategy profile is enforceable. For the special case with constant edge costs, we are furthermore able to show that Algorithm n-SEPA has polynomial running time. This has the following consequence for the *efficient* computation of a cost sharing method having low PoS: If there is an efficient (approximation) algorithm which computes a strategy profile with cost $\alpha \cdot$ OPT, where OPT denotes the cost of an optimal strategy profile and $\alpha \geq 1$, we can use the computed strategy profile as input for Algorithm n-SEPA, and get an enforceable strategy profile with cost at most $\alpha \cdot$ OPT in polynomial time. A cost sharing method enforcing this profile (which is also computed by Algorithm n-SEPA) thus has PoS at most α. This way, for the special case with constant edge costs, Algorithm n-SEPA can be seen as a *black-box* reducing the efficient computation of a (budget-balanced, stable and separable) cost sharing method with low PoS to the design of an efficient algorithm for the underlying optimization problem with low approximation guarantee. In general, computing an optimal strategy profile is $\mathcal{NP}$-hard, since the *uncapacitated facility location problem*, which is known to be $\mathcal{NP}$-hard to solve, is a special case. But for the special case of constant edge costs and *zero* delays in *undirected* n-series-parallel graphs, we show (by using a result of Bateni et al. [12]), that it is possible to compute an optimal strategy profile, and thus a cost sharing method with PoS equal to 1, in polynomial time. Finally, we show that the class of undirected *generalized series-parallel* graphs is not efficient for constant edge costs and zero delays (and $n \geq 3$ players), since there is an instance

---

[5] An edge cost function c_e is *discrete-concave* if it satisfies $c_e(x+\delta) - c_e(x) \geq c_e(y+\delta) - c_e(y)$ for all $x, y, \delta \in \mathbb{N}$ with $x \leq y$.

with an undirected generalized series-parallel graph, constant edge costs and zero delays, such that no optimal strategy profile is enforceable.

3.1.3 Discussion of the Model and Related Work

In this subsection, we discuss the model of network cost sharing games which is used in this thesis, as well as related work in the context of cost sharing games.

Cost sharing games have attracted a lot of attention in the literature. Traditionally, it has been studied in the context of cooperative game theory. In a cooperative cost sharing game (with transferable costs), there is a set of players N, and a cost value $C(S) \in \mathbb{R}$ (usually nonnegative) for each coalition of players $S \subseteq N$. The value $C(S)$ usually represents the minimum cost that the coalition S can guarantee for itself. In the context of a network game as considered in this thesis, that is, each player corresponds to a pair of vertices which she wants to be connected by at least one path, and the cost of the resulting network needs to be shared by the players, the cost of a coalition $S \subseteq N$ can be defined as the minimum cost (edge costs plus delays) of a subgraph of G containing an (s_i, t_i)-path for each $i \in S$. The goal in a cooperative cost sharing game is to find a vector of (usually nonnegative) cost shares $\xi \in \mathbb{R}^{|N|}$ for the grand coalition N which is in the *core*, that is, $\sum_{i \in N} \xi_i = C(N)$, and $\sum_{i \in S} \xi_i \leq C(S)$ for all coalitions $S \subseteq N$. Given cost shares which are in the core, no coalition has an incentive to deviate from the grand coalition (and build, for example, a network on their own). Cooperative cost sharing games have been studied over the last decades for a variety of combinatorial optimization problems, e.g. minimum spanning tree [20, 55], Steiner tree [56, 90, 102], facility location [53], vertex cover and coloring [34], and many more. See also [74] and [88] for two surveys about cooperative cost sharing games. A problem with this approach (cf. [71]) is the strong abstraction of the cost shares from the underlying optimization problem. Players only get one global cost share, and cannot decide which edges (resources, etc.) are used, and for which they spend their money on. When studying the incentives of selfish players in large settings, we need a more detailed, and *strategic* analysis of cost sharing. One such attempt is given by the following model, which we explain in the context of a network game, and which is sometimes referred to as *arbitrary cost sharing*.

In a (noncooperative) network cost sharing game with arbitrary sharing, each player i chooses additionally to her path P_i a payment vector $p_i = (p_{i,e})_{e \in E}$ which specifies the amount that player i is willing to pay for each edge $e \in E$. An edge $e \in E$ is called *bought* if $\sum_{i \in N} p_{i,e} \geq c_e(P)$, and the cost of player i

is $\sum_{e \in E} p_{i,e} + \sum_{e \in P_i} d_{i,e}$ if all edges in P_i are bought, and ∞ else.[6] Existence, quality and computability of PNE in games with arbitrary sharing have been studied for many combinatorial optimization problems, for example matroid games with nondecreasing, discrete-concave costs [63], facility location and covering games with fixed [23] and nondecreasing, discrete-concave costs [70], network games with fixed costs [6, 7, 10, 69, 72], and many more. In [71], *strong* PNE (no coalition can decrease the cost of each member of the coalition) are analyzed for various combinatorial optimization problems. In case that there is very little control over players, arbitrary cost sharing is a suitable model, since the players themselves decide how much they want to pay. In contrast, we assume in this chapter of the thesis that the cost shares are determined by a *central authority*. In this setting, budget-balanced, stable and separable cost sharing methods constitute a meaningful design space. We will now discuss these properties in some more detail.

Budget-balance, that is, the cost of an edge is exactly distributed to the users of that edge, is a very natural assumption (nevertheless, there are also approaches which relax this property, for example arbitrary sharing games). Stability requires that the induced game has at least one PNE. Although mixed-strategy equilibria always exist, their pure counterparts are often preferred, since there are many practical applications in which mixed strategies do not have a reasonable interpretation (for further discussion on mixed strategy equilibria, see for example Section 3.2 in [93]). The third, and certainly most debatable property, is separability. It means that for any two profiles P, P', the cost shares on edge e are the same if the set of players using e remains unchanged. First note that in particular, the cost sharing methods that we consider here assign cost shares per used edge, instead of a single, global cost share for each player (as it is the case for cooperative cost sharing games). When players jointly design a large network, it is more desirable to provide algorithms and methods that specify which agent needs to pay how much for each edge. Furthermore, a cost sharing method which is separable can distribute the cost of an edge in a *decentralized and local* fashion, i.e., it does not have to know any costs, or the user set, of other edges. In particular in large networks, this is a clear advantage over non-separable cost sharing methods, in which all this could be necessary.

A well-known budget-balanced and separable cost sharing method is Shapley cost sharing, and a lot of research focuses on this method. Besides obvious advantages (easy to understand and to compute, considered as "fair"), it has many further desirable properties. For example, in a network cost sharing game with delays which

---

[6]Note that for games with constant edge costs and zero delays, one usually drops the explicit path P_i from the strategy of a player. Thus, each player $i \in N$ only specifies payments $p_{i,e}$ for each $e \in E$. Then the private cost of player i is $\sum_{e \in E} p_{i,e}$ if the payments suffice to buy at least one (s_i, t_i)-path, and ∞ otherwise.

are not player-specific ($d_{i,e} = d_e$ for each $i \in N, e \in E$), a PNE always exists, and best response dynamics lead to a PNE (since the game constitutes a congestion game [9, 91, 97]). This even holds if the delays are not fixed, but the delay of an edge depends on the number of users of this edge. But as already mentioned in Section 3.1, an optimal strategy profile may not be a PNE. There even may be an equilibrium with cost n times the cost of an optimal solution, see Anshelevich et al. [9]. But it is shown in the same work that for nondecreasing, discrete-concave edge costs and fixed, player-independent delays, there always exists a PNE with cost at most the nth harmonic number H_n times the cost of an optimal solution. For directed graphs, this bound is tight (thus the worst-case PoS is H_n), since there is an instance with a directed graph, constant edge costs and zero delays having PoS equal to H_n. But for undirected graphs, the H_n upper bound on the worst-case PoS is not tight. The exact worst-case PoS of network cost sharing games with an undirected graph, constant edge costs and zero delays is a notorious open problem. The current state of the art is that the worst-case PoS is (approximately) between 2.245 and $H_{n/2}$, where the upper bound requires n to be "large enough" (see [18, 86]). Improved bounds are known for special cases of network cost sharing games. For $n = 2$ players, the worst-case PoS is known to be exactly $4/3$ (see [9, 27] and Example 3.1.2). For $n = 3$ players, it is between 1.571 and 1.634 (see [17, 37]). If all players share one source (sometimes called a *multicast game*), Li [82] showed an upper bound of $O(\frac{\log n}{\log \log n})$. If additionally each other node of the graph is the sink of a player (a *broadcast game*), or the given graph has a special bipartite structure, the worst-case PoS is $O(1)$ (see [19] and [47], respectively). If the given graph is a simple cycle, the worst-case PoS is exactly $3/2$ [41]. Besides the fact that equilibria may be expensive, they are in general also difficult to find: Computing a PNE is PLS-complete [101], and although best response dynamics always lead to a PNE, it may take exponential time [9, 101]. For further works analyzing cost sharing based on the Shapley value, see [24, 49–51, 54, 58, 59, 78, 79, 98], and many more.

More general cost sharing methods have been studied in various contexts, for example scheduling games [11, 22, 28, 44], matroid games [60, 67, 103] or network design games [25, 29, 30, 60]. Particularly, the papers of Harks and von Falkenhausen [67, 103], Chen et al. [25] and Harks et al. [60] consider the design of budget-balanced, stable and separable cost sharing methods. Harks and von Falkenhausen [67, 103] analyzed the design of cost sharing methods in games where the players' strategies are bases of matroids. In particular, they derived several tight bounds on the worst-case PoS and PoA, depending on the generality of allowed cost functions and strategy spaces. Chen et al. [25] considered the design of cost sharing methods in network cost sharing games with constant edge cost functions and zero delays. For undirected graphs, they derived (almost) tight bounds for the worst-case PoA of an optimal cost sharing method. For directed graphs, they focused on the PoS and showed that if all players have the same sink, there always exists an enforceable optimal strategy profile. For general instances in directed graphs, the worst-case PoS achieved by a best cost sharing method lies in $[3/2, H_n]$ (where H_n is again the nth harmonic number). Furthermore, Chen et al. considered the design of *uniform cost sharing protocols*. Such a protocol assigns, for every possible network, a budget-balanced, stable and separable cost sharing method, where an edge's cost sharing method may only depend on the edge cost and the player set, but not on other ingredients of the network. In their work, Chen et al. identified (almost) optimal uniform protocols for undirected and directed graphs. The results presented in Section 3.4 of this thesis are part of Harks et al. [60], where we studied the design of budget-balanced, stable and separable cost sharing methods that additionally are polynomial time computable. Besides the results from Section 3.4, we devised efficient black-box reductions to the underlying combinatorial optimization problems for single-source network cost sharing games with constant edge cost functions and zero delays, as well as matroidal set systems with subadditive costs and delays. To conclude our discussion of related work, we want to mention that PNE of arbitrary sharing games can easily be used to obtain budget-balanced, stable and separable cost sharing methods having the same PNE (this is observed in [25, 60]). This implies existence of budget-balanced, stable and separable cost sharing methods having PoS equal to 1 for a variety of classes of games, see for example [6, 7, 10, 23, 63, 69–71]. However, most of these results are computationally not tractable, meaning that there probably is no efficient algorithm which computes an optimal PNE.

3.2   An LP-Characterization of Enforceability

In this section, we present a characterization of enforceable strategy profiles. Similar results are used in [10] for *arbitrary cost sharing games*[7], and in [67, 103] for models in which the strategies are bases of matroids. In [62], we formally stated the characterization for enforceable *Steiner forests* (strategy profiles corresponding to cycle-free subgraphs) of undirected graphs with constant edge costs and zero delays, but the characterization, as well as its proof, continues to hold for the more general setting considered in this thesis.

The characterization is in terms of an optimal solution of a linear program. Roughly speaking, the linear program is used to compute the maximum amount

---

[7]See Subsection 3.1.3 for more information about arbitrary cost sharing.

that the players are "willing to pay" for the shareable costs of the used edges. If and only if this amount is enough to completely pay the shareable edge costs of all used edges, the corresponding strategy profile is enforceable. We now formally describe the linear program $LP(P)$ which is used to decide whether the strategy profile $P = (P_1, \ldots, P_n)$ is enforceable or not. For each player $i \in N$ and each edge $e \in P_i$, there is a nonnegative variable $\xi_{i,e}$ which describes the amount of the shareable edge cost that player i pays for the edge e in her path P_i. We also call $\xi_{i,e}$ the *cost share of player i for e*. The objective function is given by the sum of all variables, i.e., the joint amount that the players pay for the shareable edge costs. Besides the nonnegativity of the cost shares, we have two types of inequality constraints. The first is a relaxed budget-balance property: For each used edge $e \in E$, i.e., with $N_e(P) \neq \emptyset$, the sum of cost shares that the users of e pay for e is upper-bounded by the shareable edge cost $c_e(P)$ of e. The second type of constraints stems from the Nash equilibrium conditions for P, combined with the separability property for cost sharing methods: For each player i and each alternative path P_i' of this player, the sum of cost shares and delays that player i pays for the edges which she uses in P_i, but not in P_i', is upper-bounded by the sum of the *complete* shareable edge costs and delays for edges which she uses in P_i', but not in P_i. Edges used in P_i *and* in P_i' do not occur here, since the user sets of these edges do not change if player i changes her strategy from P_i to P_i', which implies that the cost shares induced by a *separable* cost sharing method for these edges also remain unchanged.

Altogether, we obtain the following linear program $LP(P)$ (recall that (P_i', P_{-i}) denotes the strategy profile resulting from P if player i changes her strategy from P_i to P_i'):

$$LP(P) \quad \max \sum_{e \in E: N_e(P) \neq \emptyset} \sum_{i \in N_e(P)} \xi_{i,e}$$

$$\text{s.t.:} \quad \sum_{i \in N_e(P)} \xi_{i,e} \leq c_e(P) \qquad\qquad \forall e \in E : N_e(P) \neq \emptyset \tag{1}$$

$$\sum_{e \in P_i \setminus P_i'} \left(\xi_{i,e} + d_{i,e} \right) \leq \sum_{e \in P_i' \setminus P_i} \left(c_e((P_i', P_{-i})) + d_{i,e} \right) \qquad \forall i \in N \; \forall P_i' \in \mathcal{P}_i \tag{2}$$

$$\xi_{i,e} \geq 0 \qquad\qquad\qquad\qquad \forall i \in N \; \forall e \in P_i \tag{3}$$

Note that the objective function of LP(P) is bounded, since for any feasible solution $(\xi_{i,e})_{i \in N, e \in P_i}$, property (1) of LP($P$) yields

$$\sum_{e \in E: N_e(P) \neq \emptyset} \sum_{i \in N_e(P)} \xi_{i,e} \leq \sum_{e \in E: N_e(P) \neq \emptyset} c_e(P). \tag{3.1}$$

Furthermore note that LP(P) can be infeasible in the presence of positive delays, whereas for zero delays, setting all variables to zero yields a feasible solution.

The enforceability of strategy profiles can now be characterized as follows:

Theorem 3.2.1 *A strategy profile P in a network $\mathcal{N}$ is enforceable if and only if there is an optimal solution $(\xi_{i,e})_{i \in N, e \in P_i}$ for LP(P) with*

$$\sum_{i \in N_e(P)} \xi_{i,e} = c_e(P) \; \forall e \in E : N_e(P) \neq \emptyset. \tag{BB}$$

Furthermore, given an optimal solution $(\xi_{i,e})_{i \in N, e \in P_i}$ for LP(P) with (BB), the cost sharing method $\xi = (\xi_e)_{e \in E}$ (defined below) is budget-balanced, separable and enforces P.

For each $e \in E$, the cost sharing method ξ_e assigns for each strategy profile $P' = (P'_1, \ldots, P'_n)$ with $N_e(P') \neq \emptyset$ and for each $i \in N_e(P')$ the following nonnegative cost shares:

$$\xi_{i,e}(P') := \begin{cases} \xi_{i,e}, & \text{if } N_e(P') = N_e(P), \\ c_e(P'), & \text{if } N_e(P') \setminus N_e(P) \neq \emptyset \text{ and } i = \min(N_e(P') \setminus N_e(P)), \\ c_e(P'), & \text{if } N_e(P') \subsetneq N_e(P) \text{ and } i = \min N_e(P'), \\ 0, & \text{else.} \end{cases}$$

Before presenting a proof of Theorem 3.2.1, we would like to mention the following observations regarding solutions for LP(P), which follow immediately from condition (1) of LP(P) and inequality (3.1):

Observation 3.2.2

1. A feasible solution for LP(P) with (BB) is optimal.
2. A feasible solution for LP(P) fulfills (BB) if and only if the objective function value equals the sum of edge costs over all edges used in P.

3. There is an optimal solution for LP(P) with (BB) if and only if every optimal
   solution fulfills (BB).

Proof of Theorem 3.2.1 We first assume that there is an optimal solution
$(\xi_{i,e})_{i\in N, e\in P_i}$ for LP(P) with (BB). We show that the cost sharing method $\xi =$
$(\xi_e)_{e\in E}$ given in the theorem statement is budget-balanced, separable and enforces
P, showing that P is enforceable. Since property (BB) is fulfilled for the LP(P)-
solution $(\xi_{i,e})_{i\in N, e\in P_i}$, the cost sharing method ξ is budget-balanced. Furthermore,
ξ is separable, because the defined cost shares for an edge only depend on the user
set of that edge. It remains to show that P is a PNE of the game $(\mathcal{N}, \xi)$, i.e., we
have to show that there is no player who can strictly decrease her cost by a unilateral deviation. Thus let $i \in N$ and $P_i' \in \mathcal{P}_i$, and write $P' := (P_i', P_{-i})$. We need
to show that $\xi_i(P) \le \xi_i(P')$ holds. Recall that, by definition of the players' cost
functions, $\xi_i(P) = \sum_{e\in P_i}(\xi_{i,e}(P) + d_{i,e})$ and $\xi_i(P') = \sum_{e\in P_i'}(\xi_{i,e}(P') + d_{i,e})$.
For each $e \in P_i$, the definition of ξ yields $\xi_{i,e}(P) = \xi_{i,e}$. For edges in P_i', we get
$\xi_{i,e}(P') = \xi_{i,e}$ if $e \in P_i' \cap P_i$ and $\xi_{i,e}(P') = c_e(P')$ if $e \in P_i' \setminus P_i$ (due to the
first, respectively second case in the definition of $\xi_{i,e}(P')$). Using this, as well as
condition (2) of LP(P), we have

$$\xi_i(P) = \sum_{e\in P_i\setminus P_i'} \left(\xi_{i,e} + d_{i,e}\right) + \sum_{e\in P_i\cap P_i'} \left(\xi_{i,e} + d_{i,e}\right)$$

$$\le \sum_{e\in P_i'\setminus P_i} \left(c_e(P') + d_{i,e}\right) + \sum_{e\in P_i'\cap P_i} \left(\xi_{i,e} + d_{i,e}\right) = \xi_i(P'),$$

as desired.

Now assume that P is enforceable, that is, there is a budget-balanced and separable cost sharing method ξ such that P is a PNE of the game $(\mathcal{N}, \xi)$. We show that we
get an optimal solution for LP(P) with property (BB) by setting $\xi_{i,e} := \xi_{i,e}(P)$ for
all $i \in N$ and $e \in P_i$. Due to the budget-balance of ξ, property (BB) obviously holds.
In particular, condition (1) of LP(P) is fulfilled. Condition (3), i.e., nonnegativity
of the variables, also holds true since the cost shares are nonnegative. It remains to
verify condition (2) of LP(P) (the optimality then follows from Observation 3.2.2).
Let $i \in N$ and $P_i' \in \mathcal{P}_i$. Since P is a PNE for the game $(\mathcal{N}, \xi)$, we get

$$\sum_{e\in P_i} \left(\xi_{i,e}(P) + d_{i,e}\right) = \xi_i(P) \le \xi_i((P_i', P_{-i})) = \sum_{e\in P_i'} \left(\xi_{i,e}((P_i', P_{-i})) + d_{i,e}\right),$$

which is equivalent to

$$\sum_{e \in P_i \setminus P_i'} (\xi_{i,e}(P) + d_{i,e}) + \sum_{e \in P_i \cap P_i'} (\xi_{i,e}(P) + d_{i,e})$$

$$\leq \sum_{e \in P_i' \setminus P_i} (\xi_{i,e}((P_i', P_{-i})) + d_{i,e}) + \sum_{e \in P_i \cap P_i'} (\xi_{i,e}((P_i', P_{-i})) + d_{i,e}).$$

Using that ξ is separable, we get $\xi_{i,e}(P) = \xi_{i,e}((P_i', P_{-i}))$ for each $e \in P_i \cap P_i'$, thus the inequality above yields

$$\sum_{e \in P_i \setminus P_i'} (\xi_{i,e}(P) + d_{i,e}) \leq \sum_{e \in P_i' \setminus P_i} (\xi_{i,e}((P_i', P_{-i})) + d_{i,e}). \tag{3.2}$$

Note that $\xi_{i,e}((P_i', P_{-i})) \leq c_e((P_i', P_{-i}))$ holds for each edge $e \in P_i' \setminus P_i$ due to nonnegativity and budget-balance of the cost shares. Using this, the definition of the LP(P)-variables ($\xi_{i,e} = \xi_{i,e}(P)$ for all $e \in P_i$), and (3.2), condition (2) follows, completing the proof:

$$\sum_{e \in P_i \setminus P_i'} (\xi_{i,e} + d_{i,e}) = \sum_{e \in P_i \setminus P_i'} (\xi_{i,e}(P) + d_{i,e})$$

$$\leq \sum_{e \in P_i' \setminus P_i} (\xi_{i,e}((P_i', P_{-i})) + d_{i,e})$$

$$\leq \sum_{e \in P_i' \setminus P_i} (c_e((P_i', P_{-i})) + d_{i,e})$$

$\square$

At the end of this section, we want to emphasize that LP(P) has $O(n \cdot |E|)$ variables, but it may have *exponentially* many constraints, since the number of paths connecting two specific nodes can be exponential in n. Therefore it is not clear whether LP(P) can be solved in polynomial time (with respect to the encoding length of the network $\mathcal{N} = (G, (s_i, t_i)_{i \in N}, c, d)$). Nevertheless, the presented characterization is very useful to derive our results about (strong) efficiency. Furthermore, we show in Section 3.4 that for networks having a special series-parallel structure and constant edge cost functions, LP(P) can be solved efficiently. Finally, Theorem 3.2.1 is constructive (in the sense that if P is enforceable, it also provides a cost sharing method ξ enforcing P), and the provided cost sharing method ξ is polynomial-space representable (meaning that additional to the network $\mathcal{N}$, we only have to store the solution $(\xi_{i,e})_{i \in N, e \in P_i}$ of LP(P)). Note that in principle, a cost sharing method needs to specify a sharing of the costs for each of the (possibly exponentially many) strategy profiles.

3.3 Results for Two Player Games

In this section, we consider network cost sharing games with *two* players (denoted by Player 1 and Player 2) in *undirected* graphs with *constant edge cost functions* and *zero delays*. Our main result is a complete characterization of (strong) efficiency for this setting. The characterization is in terms of forbidden subgraphs: We present a list of graphs, called *Bad Configurations*, having the property that if $(G, (s_1, t_1), (s_2, t_2))$ does not contain any of these graphs as a subgraph, then $(G, (s_1, t_1), (s_2, t_2))$ is strongly efficient for constant edge costs and zero delays. Conversely, we can show that efficiency implies that $(G, (s_1, t_1), (s_2, t_2))$ does not contain a Bad Configuration. Since strong efficiency obviously implies efficiency, we get the following result:

Theorem 3.3.1 (characterization of (strong) efficiency). *For an undirected graph G with source-sink pairs (s_1, t_1) and (s_2, t_2), the following three statements are equivalent:*

(1) $(G, (s_1, t_1), (s_2, t_2))$ *does not contain a subgraph which is a Bad Configuration (see Definition 3.3.5).*

(2) $(G, (s_1, t_1), (s_2, t_2))$ *is strongly efficient for constant edge costs and zero delays.*

(3) $(G, (s_1, t_1), (s_2, t_2))$ *is efficient for constant edge costs and zero delays.*

For notational convenience, we mostly omit the specification *for constant edge costs and zero delays* in this section, i.e., whenever we speak of (strong) efficiency, we mean with respect to constant edge costs and zero delays. Furthermore, slightly abusing notation, we write $c_e \geq 0$ for the constant value of the edge cost function c_e of edge $e \in E$. The rest of this section is organized as follows. In Subsection 3.3.1, we highlight the connection between strategy profiles and *Steiner forests*. This will become useful in the proof of Theorem 3.3.1. Then, in Subsection 3.3.2, we give some intuition behind the Bad Configurations, and present the complete list of these graphs. After this, we give a proof sketch of Theorem 3.3.1 in Subsection 3.3.3. The formal proofs of the two main directions of the theorem are contained in Subsection 3.3.4 and Subsection 3.3.5. Concluding, we discuss our result in Subsection 3.3.6, in particular according to possible implications and generalizations.

3.3.1   Connection to Steiner forests

In this subsection, we want to highlight the connection between strategy profiles and *Steiner forests*. This will be used in the proof of Theorem 3.3.1.

Given an undirected graph $G = (V, E)$ and source-sink pairs (s_i, t_i) for $i \in N$, a *Steiner forest* $F \subseteq E$ is a subset of the edges of G, such that the induced subgraph of G is cycle-free and contains an (s_i, t_i)-path for each $i \in N$. Equivalently, the induced subgraph contains a unique (s_i, t_i)-path P_i for each $i \in N$. Thus, there is a unique strategy profile (namely $P = (P_1, \ldots, P_n)$) corresponding to F. Note that this way, one can also speak of an *enforceable Steiner forest*, namely if the corresponding strategy profile is enforceable. On the other hand, given an arbitrary strategy profile P, the set of used edges does not necessarily induce a cycle-free subgraph. But obviously, by deleting suitable edges from the set of used edges, one obtains a Steiner forest.

Now assume that we have constant edge costs $c = (c_e)_{e \in E}$. The cost of a Steiner forest F is then defined as the sum of edge costs over edges in F, i.e., $C(F) := \sum_{e \in F} c_e$. An *optimal* Steiner forest is a cost-minimal Steiner forest. Since we have no delays, the cost of the corresponding strategy profile equals the cost of the Steiner forest. On the other hand, using that edge costs are nonnegative, each Steiner forest obtained from a strategy profile (possibly by deleting suitable edges) has cost less than or equal to the cost of the strategy profile. Thus the cost of an optimal Steiner forest equals the cost of an optimal strategy profile, and consequently, the strategy profile corresponding to an optimal Steiner forest is optimal.

In light of (strong) efficiency, it is also important to analyze the set of *all* optimal strategy profiles. In particular it is useful to know, under which condition this set "equals" the set of optimal Steiner forests (in the sense that we get all optimal strategy profiles by taking the profiles corresponding to optimal Steiner forests). To this end, assume that we have an optimal strategy profile P such that the used edges do not constitute a Steiner forest. This means that the induced subgraph contains at least one cycle, and for each edge e which is contained in any cycle, we get $c_e = 0$ (due to the optimality of the strategy profile). In particular this shows the following observation, which will become useful in the proof of $(3) \Rightarrow (1)$ of Theorem 3.3.1.

Observation 3.3.2 Let $\mathcal{N} = (G, (s_i, t_i)_{i \in \{1,2\}}, c, d)$ be a network with an undirected graph $G = (V, E)$, constant edge cost functions $c = (c_e)_{e \in E}$ and zero delays. If G does not contain a cycle C such that $c_e = 0$ for all $e \in C$, then the set of optimal strategy profiles corresponds to the set of optimal Steiner forests.

In the rest of this subsection, we show that the existence of an optimal, but not enforceable strategy profile implies the existence of an optimal, not enforceable *Steiner forest*. We will use this in the proof of (1) $\Rightarrow$ (2) of Theorem 3.3.1.

Lemma 3.3.3 *Let $\mathcal{N} = (G, (s_i, t_i)_{i \in \{1,2\}}, c, d)$ be a network with an undirected graph $G = (V, E)$, constant edge cost functions $c = (c_e)_{e \in E}$ and zero delays. If there is an optimal strategy profile which is not enforceable, then there also exists an optimal Steiner forest which is not enforceable.*

Proof Assume that there exists an optimal strategy profile which is not enforceable. Among all such strategy profiles, choose $P = (P_1, P_2)$ such that the number of used edges is minimal. We show that the edges which are used in P constitute a Steiner forest. Assume, by contradiction, that this does not hold. We now construct a strategy profile P' which is optimal, not enforceable, and has less used edges than P, contradicting our choice of P and completing the proof. Figure 3.3 illustrates the following argumentation.

Let C be a cycle in the subgraph induced by P. Since P_1 does not contain a cycle, there is an edge $e \in C$ such that Player 1 does not use edge e. Let C_2 be the subpath of C which is defined by the following properties: The edge e is contained in C_2, and the two endvertices u and v of C_2 are contained in P_1, but no other vertex of C_2 is contained in P_1. Note that C_2 exists (since P_1 contains at least one edge of C) and is a subpath of P_2. Let P_{uv} be the subpath of P_1 with endvertices u and v. We now show that the strategy profile $P' = (P_1', P_2)$ resulting from P if Player 1 uses C_2 instead of the subpath P_{uv} has the desired properties, i.e., is optimal, not enforceable, and has less used edges than P. First note that the union of P_{uv} and C_2 is a cycle C'. Since P_2 does not contain this cycle, P_{uv} contains at least one edge which is not used by Player 2. This shows that the edges which are used in P' constitute a proper subset of the edges which are used in P. Therefore P' is optimal and has less used edges than P. It remains to show that P' is not enforceable. To this end, note that all edges $e \in C'$ have cost $c_e = 0$ (due to the optimality of P). Thus, Player 1 substituted some edges with cost zero by other edges having cost zero. We now argue that such a change does not influence enforceability.

To see this, consider an arbitrary strategy profile P, a feasible solution $(\xi_{i,e})_{i \in N, e \in P_i}$ for LP(P) and a used edge f with $c_f = 0$. The properties (1) and (3) of LP(P) immediately imply $\xi_{i,f} = 0$ for each player i using edge f in her path. Furthermore, edge f is always completely paid. Thus, we can "delete" all edges having cost zero from LP(P) (resulting in the linear program LP(P)$_{>0}$ below) and characterize the enforceability of P by using the new program: P is enforceable if and only if there is an optimal solution $(\xi_{i,e})_{i \in N, e \in P_i : c_e > 0}$ for LP(P)$_{>0}$ such that all used edges e with $c_e > 0$ are completely paid. Now, if some player i substitutes some edges having cost zero by other edges with cost zero, resulting in the strategy profile P', we get that $\{e \in P_i : c_e > 0\} = \{e \in P'_i : c_e > 0\}$ holds, and thus the programs LP(P)$_{>0}$ and LP(P')$_{>0}$ are exactly the same linear programs. This shows that P' is enforceable if and only if P is enforceable.

For the specific strategy profiles P and P' considered in this proof, the above argumentation shows that P' is not enforceable, since P is not enforceable.

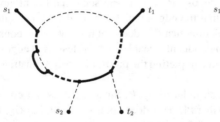

The profile P with P_1 thick, P_2 dashed.

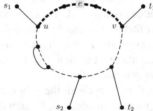

The cycle C (dashed) with subpath C_2 (thick).

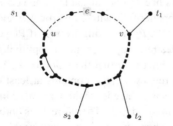

The cycle C' (dashed) with subpath P_{uv} (thick).

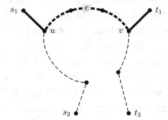

The profile P' with P'_1 thick, P_2 dashed.

Fig. 3.3 Illustration for the proof of Lemma 3.3.3

$$\text{LP}(P)_{>0} \quad \max \sum_{e \in E: N_e(P) \neq \emptyset,\, c_e > 0} \sum_{i \in N_e(P)} \xi_{i,e}$$

$$\text{s.t.:} \quad \sum_{i \in N_e(P)} \xi_{i,e} \leq c_e \qquad\qquad \forall e \in E : N_e(P) \neq \emptyset, c_e > 0$$

$$\sum_{e \in P_i \setminus P_i',\, c_e > 0} \xi_{i,e} \leq \sum_{e \in P_i' \setminus P_i,\, c_e > 0} c_e \qquad \forall i \in N \;\forall P_i' \in \mathcal{P}_i$$

$$\xi_{i,e} \geq 0 \qquad\qquad\qquad \forall i \in N \;\forall e \in P_i : c_e > 0$$

$\square$

3.3.2 The Set of Bad Configurations and some Intuition

The aim of this subsection is to provide some intuition behind the *Bad Configurations*, the subgraphs that we need to exclude for (strong) efficiency, and to present the complete list of these graphs.

For a start, we consider an example of a graph which is not efficient.

Proposition 3.3.4 *Let* $(G, (s_1, t_1), (s_2, t_2))$ *be the undirected graph with source-sink pairs given in Figure 3.4.[8] Then* $(G, (s_1, t_1), (s_2, t_2))$ *is not efficient for constant edge costs and zero delays.*

Fig. 3.4 An example of a non-efficient graph. The edges are labelled with their name, followed by their (constant) cost. The thick edges build the unique optimal Steiner forest

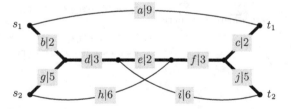

Proof Let $c = (c_e)_{e \in E}$ be the constant edge costs given in Figure 3.4, and define all delays as zero $d = (d_{i,e} := 0)_{i \in N, e \in E}$. By Definition 3.1.4, it suffices to show that no optimal strategy profile of the network $\mathcal{N} := (G, (s_1, t_1), (s_2, t_2), c, d)$

---

[8]Note that there is a very similar directed graph which is not efficient for constant edge costs and zero delays, see Chen et al. [25].

is enforceable. Note that since the given edge costs are strictly positive, it follows from Observation 3.3.2 that we only need to show that there is no optimal Steiner forest which is enforceable. One can easily verify by comparing all 19 possible Steiner forests that there is a unique optimal Steiner forest, given by the (thick) edges $\{b, c, d, e, f, g, j\}$. The corresponding strategies of the two players are $P_1 = \{b, d, e, f, c\}$ and $P_2 = \{g, d, e, f, j\}$. Write OPT $= (P_1, P_2)$ for the corresponding (unique) optimal strategy profile. It remains to show that OPT is not enforceable. To this end, consider the linear program LP(OPT) introduced in Section 3.2. Note that LP(OPT) is feasible, since setting all variables to zero yields a feasible solution for the case of zero delays. Consider an arbitrary, feasible solution $(\xi_{i,e})_{i \in N, e \in P_i}$ of LP(OPT). We will show that the objective function value is strictly smaller than the sum of edge costs over all used edges, i.e., the cost $C(\text{OPT})$ of OPT, which implies by Theorem 3.2.1 and Observation 3.2.2 that OPT is not enforceable. By considering the paths $P_1' = \{a\}$, $P_2' = \{h, f, j\}$ and $P_2'' = \{g, d, i\}$, we get from (2) of LP(OPT):

$$\xi_{1,b} + \xi_{1,d} + \xi_{1,e} + \xi_{1,f} + \xi_{1,c} \leq 9$$
$$\xi_{2,g} + \xi_{2,d} + \xi_{2,e} \leq 6$$
$$\xi_{2,e} + \xi_{2,f} + \xi_{2,j} \leq 6$$

Together with the nonnegativity of the variables ((3) of LP(OPT)), this yields the desired inequality, completing the proof:

$$\sum_{e \in E: N_e(\text{OPT}) \neq \emptyset} \sum_{i \in N_e(\text{OPT})} \xi_{i,e} = \xi_{1,b} + \xi_{1,c} + \xi_{1,d} + \xi_{2,d} + \xi_{1,e} + \xi_{2,e} + \xi_{1,f} + \xi_{2,f} + \xi_{2,g} + \xi_{2,j}$$
$$\leq \xi_{1,b} + \xi_{1,c} + \xi_{1,d} + \xi_{2,d} + \xi_{1,e} + 2\xi_{2,e} + \xi_{1,f} + \xi_{2,f} + \xi_{2,g} + \xi_{2,j}$$
$$\leq 9 + 6 + 6 = 21 < 22 = C(\text{OPT})$$

$\square$

Recall that we want to find a (short) list of graphs, called *Bad Configurations*, that completely characterizes the class of (strongly) efficient graphs. We have already seen that the graph displayed in Figure 3.4 is not efficient and in fact it constitutes one Bad Configuration. We now give some intuition why we achieve a strongly efficient graph, if the graph of Figure 3.4, or the other quite similar Bad Configurations, are excluded as subgraphs. Or, equivalently, why a not strongly efficient graph needs to contain a Bad Configuration as a subgraph. All details, and formal definitions of the used terminology, can be found in Subsection 3.3.5.

Let $(G, (s_1, t_1), (s_2, t_2))$ be not strongly efficient. That is, there are constant edge costs $c = (c_e)_{e \in E}$ and an optimal strategy profile P which is not enforceable. By Lemma 3.3.3, we can assume that P corresponds to a Steiner forest F. Let $(\xi_{i,e})_{i \in N, e \in P_i}$ be an optimal LP(P)-solution. As in Section 3.2, we call the values $\xi_{i,e}$ *cost shares*. Since F is not enforceable, there is an edge $e \in F$ with $\sum_{i \in N_e(P)} \xi_{i,e} < c_e$. We say that e is *not completely paid*. Figure 3.5a shows a "prototype" of F, in which each player first uses some edges on her own (the *left part* of F), then both players share some edges (the *middle part* of F), and in the end they separate again (the *right part* of F). Note that the displayed lines may consist of more than one edge. By requiring more structure for the cost shares, see Subsection 3.3.5, we can assume w.l.o.g. that the edge e (which is not completely paid) is located in the right part of F. Using the optimality of $(\xi_{i,e})_{i \in N, e \in P_i}$ for LP(P), Player 1 needs to have a path $P'_i \in \mathcal{P}_i$ with $e \notin P'_i$, such that the corresponding inequality (2) of LP(P) is tight. We call the edges in $P'_i \setminus P_i$ a *tight alternative* for e, and say that $P_i \setminus P'_i$ is *substituted* by the tight alternative. We now analyze how the tight alternative of Player 1 for e may look like. If it only substitutes edges which Player 1 uses alone, as displayed in Figure 3.5b, we get a contradiction to the optimality of F (deleting the thick edges and adding the dashed edges yields a cheaper Steiner forest). Thus, we can assume w.l.o.g. that the tight alternative for e also substitutes some *commonly used edges*. Either, the tight alternative "ends" in the commonly used subpath (called *small*, see Figure 3.5c), or it ends in the subpath leading to s_1 that Player 1 uses alone (called *large*, see Figure 3.5d). We now proceed with the case of Figure 3.5d, the other case leads to slightly different Bad Configurations. If Player 2 completely pays all commonly used edges, i.e., $\xi_{2,f} = c_f$ holds for all edges $f \in F$ used by both players, we again get a contradiction to the optimality of F (deleting the edges that Player 1 uses alone and adding the dashed edges yields a cheaper Steiner forest). Therefore, there has to be an edge f that Player 2 does not pay completely. Intuitively, Player 2 needs to have a tight alternative for f, which "prevents" her to pay more for it. Otherwise, it should be possible to change the cost shares in a way that leads to a feasible solution for LP(P) with higher objective than the given cost shares, contradicting the optimality of $(\xi_{i,e})_{i \in N, e \in P_i}$. If $\xi_{1,f} + \xi_{2,f} < c_f$ holds, it is immediately clear how to change the cost shares (namely just increase $\xi_{2,f}$ by some small enough amount). The case $\xi_{1,f} + \xi_{2,f} = c_f$ is more involved. By requiring additional structure for the cost shares, however, we can also ensure the existence of a tight alternative in this case (see Subsection 3.3.5). A similar argumentation shows that Player 2 has tight alternatives for *all* of the commonly used edges. We now analyze how these alternatives may look like. If there is one tight alternative substituting all commonly edges, or a "combination" of tight alternatives substituting all commonly used edges (both cases are displayed in Figure 3.5e), we get a cheaper Steiner forest than F by using the tight alternatives of both players. Thus, it has to be the case that the tight alternatives of Player 2 cannot be "combined" without using some commonly used edges, i.e., there are two tight alternatives as displayed in Figure 3.5f. The constructed subgraph now looks like the graph displayed in Figure 3.4. But note that the tight alternatives in Figure 3.5f are simple paths, possibly consisting of more than one edge. Therefore, it is possible that a tight alternative intersects with another tight alternative, and/or with the other player's path in F, see the Figures 3.5g and 3.5h for two examples. If there are no other than the displayed intersections, it turns out that the graph displayed in Figure 3.5g is not efficient, and thus it is a Bad Configuration, whereas the graph in Figure 3.5h is efficient and thus is not a Bad Configuration. We call the graph displayed in Figure 3.5f a *Preliminary Bad Configuration* and we need to carefully analyze all possible ways how the tight alternatives may intersect with each other and with the paths in F, leading to Bad Configurations, and *No Bad Configurations*. For formal definitions, see Definition 3.3.12, Definition 3.3.13 and Definition 3.3.14.

To conclude this subsection, we provide the complete list of Bad Configurations (BCs) in the following definition (for a more formal definition, see Definition 3.3.13). Note that the BCs either arise from the Preliminary Bad Configuration displayed in Figure 3.5f, or from an almost identical graph, where the tight alternative of Player 1 is small (compare Figure 3.5c).

Definition 3.3.5 (Bad Configurations (BCs)). In Figure 3.6, we display the graphs with designated vertices $\{u, v, w, x\}$ which we call *Bad Configurations (BCs)*, where the following points need to be taken into account (in particular according to the illustration in Figure 3.6):

- $\{u, v\}$ are source and sink of one player, and $\{w, x\}$ source and sink of the other player.
- Lines represent simple paths, and paths are pairwise internal node-disjoint.
- Solid paths have to consist of at least one edge, whereas dashed paths may consist of only one node.

As indicated in Figure 3.6, we differ between nine types of BCs, denoted by BC1a, BC1b, BC2a, BC2b, BC2c, BC2d, BC3, BC4a and BC4b.

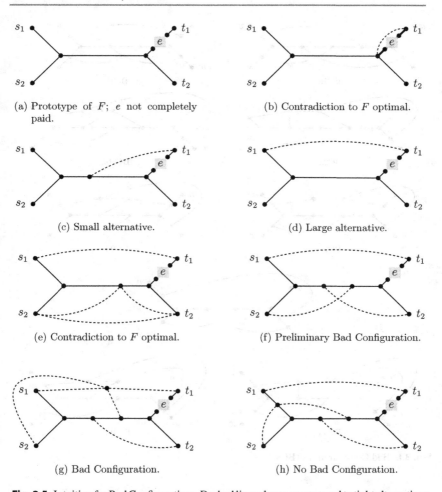

(a) Prototype of F; e not completely paid.

(b) Contradiction to F optimal.

(c) Small alternative.

(d) Large alternative.

(e) Contradiction to F optimal.

(f) Preliminary Bad Configuration.

(g) Bad Configuration.

(h) No Bad Configuration.

Fig. 3.5 Intuition for Bad Configurations. Dashed lines always correspond to tight alternatives

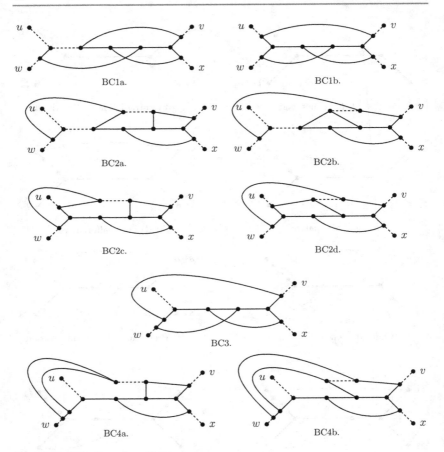

Fig. 3.6 Bad Configurations (BCs)

3.3.3 Proof Sketch of Theorem 3.3.1

In this subsection, we sketch the proof of Theorem 3.3.1. The proof proceeds in the order $(1) \Rightarrow (2) \Rightarrow (3) \Rightarrow (1)$, where $(2) \Rightarrow (3)$ follows trivially.

The direction (3) $\Rightarrow$ (1), i.e., an efficient graph does not contain a BC as a subgraph, is shown by contraposition. Thus consider a graph which contains a BC as a subgraph. We first consider this BC separately and derive costs for the edges of the BC, such that, in the subnetwork induced by the BC, no optimal strategy profile is enforceable. Then, for the remaining edges of the complete graph, we assign sufficiently high costs such that these edges are effectively deleted, meaning that they are not contained in any optimal strategy profile of the complete network. This implies that no optimal strategy profile of the complete network is enforceable, and thus the given graph is not efficient.

The direction (1) $\Rightarrow$ (2), that is, showing that every graph which does not contain any of the BCs as a subgraph is strongly efficient, is much more involved. We prove this by contradiction, thus assume there is a graph which does not contain a BC, but is not strongly efficient. By definition, there are constant edge costs so that there is an optimal, not enforceable strategy profile. Without loss of generality, we can assume that this strategy profile corresponds to a Steiner forest. We then solve the corresponding LP from Section 3.2, and since the Steiner forest is not enforceable, there exists an inequality which is not tight, corresponding to an edge that is not completely paid by the players. We use this unpaid edge to derive the existence of an alternative strategy (path) for some player with costs equal to a fraction of the currently paid cost shares (this alternative strategy corresponds to a tight inequality of the LP; the existence follows from the fact that if budget-balance is violated for an edge, the players using this edge need to have at least one alternative strategy which "prevents" them from paying more). These alternative paths are now iteratively generated until we can either argue that there exists a cheaper Steiner forest compared to the initial optimal Steiner forest (contradiction), or, there is a Bad Configuration (contradiction). Along this main approach, however, several additional ideas are required: the location of the unpaid edge leads to different subcases for which we need to use special optimal LP-solutions in order to derive the proper alternative strategies.

In the next two subsections, we present the formal proofs for (3) $\Rightarrow$ (1) and (1) $\Rightarrow$ (2).

3.3.4   (3) implies (1) of Theorem 3.3.1

In this subsection, we present a formal proof for the direction (3) $\Rightarrow$ (1) of Theorem 3.3.1, that is, if $(G, (s_1, t_1), (s_2, t_2))$ is efficient (for constant edge costs and zero delays), it does not contain a BC as a subgraph.

We show this by contraposition, thus we assume that $(G, (s_1, t_1), (s_2, t_2))$ contains a Bad Configuration and conclude that $(G, (s_1, t_1), (s_2, t_2))$ is not efficient. To this end, we need to find edge costs c, such that no optimal strategy profile of the network $(G, (s_1, t_1), (s_2, t_2), c, d)$ is enforceable (with $d_{i,e} = 0$ for $i \in N, e \in E$). Obviously, it is sufficient to find edge costs so that the optimal strategy profile is *unique* and not enforceable. We now argue why we only need to find edge cost with these properties for the BCs. To this end, let $G' = (V', E')$ a subgraph of $G = (V, E)$ that is a BC. Assume that we have edge costs $c' = (c'_{e'})_{e' \in E'}$ for the BC such that the optimal strategy profile (of the subnetwork induced by the BC) is unique and not enforceable. Denote this strategy profile by OPT, and denote its cost by $C'(\text{OPT})$. Now consider the whole graph $G = (V, E)$. For edges $e' \in E'$, i.e., edges in the BC, define $c_{e'} := c'_{e'}$, and for edges $e \in E \setminus E'$, define c_e *sufficiently large*, e.g. $c_e := C'(\text{OPT}) + 1$. Obviously, an edge $e \in E \setminus E'$ will not be contained in any optimal strategy profile, thus OPT is the unique optimal strategy profile (of the complete network) and not enforceable, as desired. In Figure 3.7, we displayed an example for the described construction.

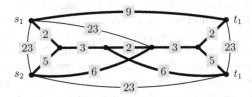

Fig. 3.7 Example for the construction in the proof of (3) $\Rightarrow$ (1). The contained BC (thick edges) is a BC1b, for which we already found edge costs with the desired properties (see Proposition 3.3.4)

To complete the proof of (3) $\Rightarrow$ (1), it remains to find edge costs for each of the BCs such that the optimal strategy profile is unique and not enforceable. In Figure 3.8, we have displayed costs for each type of BC (note that all paths with nonzero costs contain at least one edge because of the definition of the corresponding type of BC). These costs yield the desired properties (i.e., a unique optimal strategy profile OPT, and OPT is not enforceable): First note that there are no cycles having cost zero, thus the optimal strategy profiles correspond to the optimal Steiner forests (due to 3.3.2). Furthermore, there is always a unique optimal Steiner forest (given by the thick paths). Finally, the corresponding unique optimal strategy profile OPT is always not enforceable: This follows from the fact that there is always an upper bound for the objective function of LP(OPT) which is strictly smaller than the cost $C(\text{OPT})$ of OPT, and this shows that OPT is not enforceable (by Theorem 3.2.1 and Observation 3.2.2). Note that we already explained this in detail in the proof of Proposition 3.3.4 for the graph given in Figure 3.4, which is a BC1b. Additionally, we now exemplarily show the bound for BC4a. The bounds for the remaining BCs follow by an analogous argumentation, thus we now state them without further explanation: For BC1a,b and BC3, we can upper-bound the objective function by $9 + 6 + 6 = 21 < 22 = C(\text{OPT})$; for BC2a-d and BC4a,b, the objective function is bounded from above by $11 + 8 + 6 = 25 < 26 = C(\text{OPT})$.

Now turn to BC4a. In Figure 3.9, all subpaths of BC4a are labelled with their name, followed by their cost (as given in Figure 3.8). Note that for notational convenience, we have chosen $(u, v, w, x) = (s_1, t_1, s_2, t_2)$; the argumentation for the other cases is analogous. The unique optimal Steiner forest is given by the thick paths $\{a, b, c, d, e, f, g, h, i, j, k\}$. The corresponding unique optimal strategy profile OPT has cost $C(\text{OPT}) = 26$. Player 1's path in OPT is $P_1 = \{a, e, f, g, h, i\}$, and Player 2's path is $P_2 = \{b, c, d, e, f, g, j, k\}$. We now consider an arbitrary feasible solution $(\xi_{i,e})_{i \in \{1,2\}, e \in P_i}$ of LP(OPT). By considering $P_1' = \{a, d, \ell, n, o, i\}$, $P_2' = \{b, m, n, p, g, j, k\}$ and $P_2'' = \{b, c, d, e, q, k\}$, property (2) of LP(OPT) yields the following inequalities (where we write $\xi_{1,\alpha} := \sum_{e \in \alpha} \xi_{1,e}$ and $\xi_{2,\alpha} := \sum_{e \in \alpha} \xi_{2,e}$ for the sum that Player 1, respectively Player 2, pays for the edges contained in path α):

$$\xi_{1,e} + \xi_{1,f} + \xi_{1,g} + \xi_{1,h} \leq 0 + 5 + 0 + 6 = 11$$
$$\xi_{2,c} + \xi_{2,d} + \xi_{2,e} + \xi_{2,f} \leq 5 + 0 + 3 = 8$$
$$\xi_{2,f} + \xi_{2,g} + \xi_{2,j} \leq 6.$$

Using this, as well as $\xi_{i,\alpha} = 0$ for $i \in N$ and paths α with cost 0 (by properties (1) and (3) of LP(OPT)), we now get the desired upper bound on the objective function of LP(OPT), completing the proof of the direction (3) $\Rightarrow$ (1) of Theorem 3.3.1:

$$\sum_{e \in E: N_e(\text{OPT}) \neq \emptyset} \sum_{i \in N_e(\text{OPT})} \xi_{i,e} \leq \xi_{2,c} + \xi_{1,e} + \xi_{2,e} + \xi_{1,f} + 2\xi_{2,f} + \xi_{1,g} + \xi_{2,g} + \xi_{1,h} + \xi_{2,j}$$

$$\leq 11 + 8 + 6 = 25 < 26 = C(\text{OPT}).$$

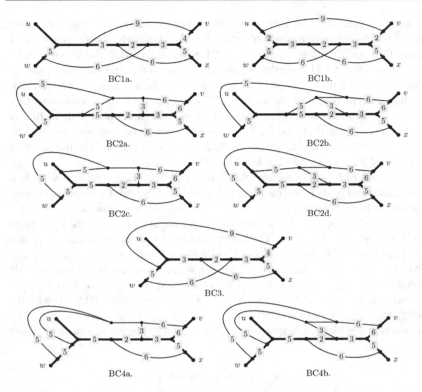

Fig. 3.8 Costs for the BCs: Paths with cost > 0 are labelled with their cost. If a path consists of more than one edge, one can choose the costs of the corresponding edges arbitrarily so that they sum up to the displayed cost on the path

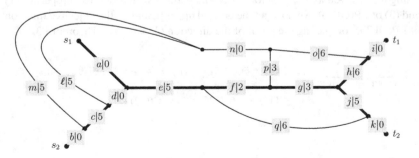

Fig. 3.9 Bounding the objective function of LP(OPT) for BC4a. Paths are labelled with their name, followed by their cost

3.3.5 (1) implies (2) of Theorem 3.3.1

In this subsection, we present a formal proof for the direction (1) $\Rightarrow$ (2) of Theorem 3.3.1, that is, if $(G, (s_1, t_1), (s_2, t_2))$ does not contain a BC as a subgraph, it is strongly efficient (for constant edge costs and zero delays). Recall that $(G, (s_1, t_1), (s_2, t_2))$ is strongly efficient for constant edge costs and zero delays if, for every vector $c = (c_e)_{e \in E}$ of constant edge costs, and zero delays $d = (d_{i,e} = 0)_{i \in N, e \in E}$, *every* optimal strategy profile of the network $(G, (s_1, t_1), (s_2, t_2), c, d)$ is enforceable. We prove this by contradiction, so assume that $(G, (s_1, t_1), (s_2, t_2))$ does not contain a BC as a subgraph, but there are constant edge costs $c = (c_e)_{e \in E}$, such that the resulting network contains an optimal strategy profile $P = (P_1, P_2)$ which is not enforceable. Due to Lemma 3.3.3, we can assume without loss of generality that the edges which are used in P are an (optimal) Steiner forest F. Furthermore, there has to be at least one edge in $P_1 \cap P_2$, i.e., which is used by both players: Otherwise, the optimality of P implies that P_1 and P_2 are shortest (s_1, t_1)-, respectively (s_2, t_2)-paths. Then, setting $\xi_{i,e} := c_e$ for $i \in N, e \in P_i$ is a feasible solution for LP(P) with property (BB), which contradicts by Theorem 3.2.1 and Observation 3.2.2 our assumption that P is not enforceable. Thus, the set of edges which are used by both players is nonempty. Furthermore, this set forms a path, since F is a Steiner forest and thus does not contain cycles. We call this set the *commonly used edges/path* or the *middle part of F*. The term *middle part* stems from the idea that the commonly used edges subdivide the players' paths into three subpaths each: For Player 1, consider P_1 from s_1 to t_1, and let u be the first, and v the last node of P_1 which is contained in the commonly used path. The three subpaths of Player 1 are then the subpaths of P_1 from s_1 to u, from u to v (the commonly used path), and from v to t_1. Considering P_2 from s_2 to t_2, we can assume w.l.o.g. that we traverse the commonly used path from u to v, that is, u is again the first, and v is again the last node of P_2 which is contained in the commonly used path: Else, that is, v is the first, and u the last node of P_2 in the commonly used path, just swap source and sink of Player 2 (in the resulting network, P is still an optimal and not enforceable strategy profile, and there is no subgraph which is a BC). The three subpaths of Player 2 are then the subpaths of P_2 from s_2 to u, from u to v, and from v to t_2. We call the subpaths from s_1 to u and from s_2 to u the *left part of F*, and the subpaths from v to t_1 and from v to t_2 the *right part of F* (see Figure 3.10a). In Figure 3.10a, we also indicate a complete ordering on the edges of F that we will use in our proof: The numbers indicate in which order we consider the subpaths; the arrows indicate increasing order within the subpaths. More formally, the edges of F are labelled with their *order* in Figure 3.10b, where ℓ_1 and r_1, respectively ℓ_2 and r_2, denote the number of edges of the subpaths of P_1, respectively P_2, in the left and right part of F, and m denotes the number of commonly used edges. Note that ℓ_1, ℓ_2, r_1 and r_2 may be zero, whereas $m \geq 1$ holds.

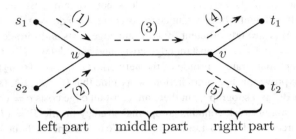

(a) Left, middle and right part of F; complete ordering on the edges of F.

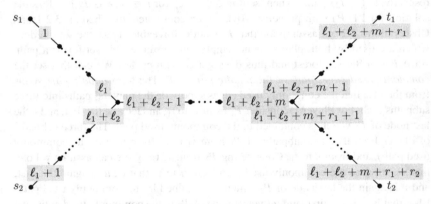

(b) Notation for the ordering of F. The edges are labelled with their order.

Fig. 3.10 Illustration for F and the used complete ordering on the edges

Since we assumed that P is not enforceable, property (BB) is violated for any optimal solution $(\xi_{i,e})_{i \in N, e \in P_i}$ of LP(P) (by Theorem 3.2.1). For our proof we consider an optimal LP(P)-solution with an additional property, given in the following definition.

Definition 3.3.6 (PTL-solution). Let $(\xi_{i,e})_{i\in N, e\in P_i}$ an optimal solution for LP(P). For a player $i \in N$, edges $e, f \in P_i$, where e has smaller order than f, and $\varepsilon > 0$, the operation PUSHLEFT(i, e, f, ε) increases the cost share $\xi_{i,e}$ of player i on edge e by ε and decreases $\xi_{i,f}$ by ε (see Figure 3.11 for illustration). If these changes yield a feasible LP(P)-solution, we call PUSHLEFT(i, e, f, ε) feasible for $(\xi_{i,e})_{i\in N, e\in P_i}$.

We call an optimal solution $(\xi_{i,e})_{i\in N, e\in P_i}$ for LP(P) *pushed to the left (PTL)*, if for any $i \in N, e, f \in E$ where e has smaller order than f, and $\varepsilon > 0$, the operation PUSHLEFT(i, e, f, ε) is not feasible for $(\xi_{i,e})_{i\in N, e\in P_i}$.

Fig. 3.11 Illustration for
PUSHLEFT(1, e, f, ε)

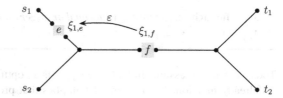

An optimal PTL-solution for LP(P) exists and can be obtained as follows. First, we introduce an ordering of the cost shares $(\xi_{i,e})_{i\in N, e\in P_i}$ which arises from the ordering of the edges of F, together with the definition that for a commonly used edge, Player 1's cost share has smaller order than Player 2's cost share. That is, a cost share $\xi_{i,e}$ has smaller order than $\xi_{j,f}$ if and only if e has smaller order than f, or $e = f$ and $i = 1, j = 2$ holds. Using this ordering, we can apply Algorithm PUSHLEFT to compute PTL-cost shares. Note that, obviously, each of the problems ($P^{j,f}$) occurring in Line 2 of Algorithm PUSHLEFT can be written as linear program. Furthermore, ($P^{j,f}$) is feasible and bounded (the current cost shares $(\xi_{i,e})_{i\in N, e\in P_i}$ are a feasible solution, and $\tilde{\xi}_{j,f}$ is upper-bounded by c_f), thus an optimal solution always exists. The correctness of Algorithm PUSHLEFT is also easy to see: Obviously, the computed cost shares $(\xi_{i,e})_{i\in N, e\in P_i}$ are optimal for LP(P), and the existence of a feasible push operation with i, e, f, ε contradicts the maximality of $\xi_{i,e}$ for ($P^{i,e}$).

Algorithm 1: PUSHLEFT

Input: Optimal solution $(\xi_{i,e})_{i\in\{1,2\},e\in P_i}$ for LP(P).

Output: Optimal PTL-solution $(\xi_{i,e})_{i\in\{1,2\},e\in P_i}$ for LP(P).

1 **foreach** *cost share* $\xi_{j,f}$ *(in increasing order)* **do**

2   Let $(\tilde{\xi}_{i,e})_{i\in\{1,2\},e\in P_i}$ be an optimal solution for the problem

$$(P^{j,f}) \quad \max \ \tilde{\xi}_{j,f}$$

$$\text{s.t.:} \ \tilde{\xi}_{i,e} = \xi_{i,e} \quad \text{for all cost shares } \xi_{i,e} \text{ with smaller order than } \xi_{j,f}$$

$$(\tilde{\xi}_{i,e})_{i\in\{1,2\},e\in P_i} \text{ optimal LP(P)-solution}$$

3   **foreach** *cost share* $\xi_{i,e}$ *with equal or larger order than* $\xi_{j,f}$ **do**

4        $\lfloor \ \xi_{i,e} \leftarrow \tilde{\xi}_{i,e}.$

Thus we can now assume that $(\xi_{i,e})_{i\in N,e\in P_i}$ is an optimal PTL-solution of LP(P). As already mentioned, Theorem 3.2.1 implies that property (BB) is not fulfilled. Let $e \in E$ with $N_e(P) \neq \emptyset$ be the *first* edge (i.e., with *smallest order*) which is not completely paid, i.e., with $\sum_{i\in N_e(P)} \xi_{i,e} < c_e$. We distinguish between the following three cases according to the "location" of e, where ord(e) denotes the order of e:

Case L: e is in the left part of F; ord(e) $\leq \ell_1 + \ell_2$.

Case M: e is in the middle part of F; $\ell_1 + \ell_2 + 1 \leq$ ord(e) $\leq \ell_1 + \ell_2 + m$.

Case R: e is in the right part of F; $\ell_1 + \ell_2 + m + 1 \leq$ ord(e).

In the following, we analyze all three cases, and derive a contradiction for each case which shows that this case cannot hold. Overall this shows that our assumption, that $(G, (s_1, t_1), (s_2, t_2))$ is not strongly efficient, is false, and completes the proof of (1) $\Rightarrow$ (2) of Theorem 3.3.1.

Each of the three Cases L, M and R is treated in a separate subsection (Case L in Subsection 3.3.5.1, Case M in Subsection 3.3.5.2 and Case R in Subsection 3.3.5.3). It turns out that the Cases L and M are quite straightforward, whereas the analysis of Case R is long and complicated (there are a lot of different subcases to analyze, and we need even more structure on the cost shares).

Before we start with the analysis of the three cases, we introduce some useful notation that will be used throughout the proof.

Notation 3.3.7

(N1) For $\alpha \in \{1, \ldots, \ell_1 + \ell_2 + m + r_1 + r_2\}$, we denote the edge of order α with e_α. In all following figures, we label the edges (except e) with their order.

(N2) For a subset of edges $S \subseteq F$, we call an edge $e_\alpha \in S$ the *smallest (largest)* edge in S, if all other edges in S have larger (smaller) order than α.

(N3) We call $P_i' \in \mathcal{P}_i$ a *tight alternative* of player i for edge e_α, if $e_\alpha \notin P_i'$ and the corresponding inequality in (2) of LP(P) is tight. Adding the edges of P_i' to P_i yields a unique cycle C with $e_\alpha \in C$. Let e_β (e_γ) be the smallest (largest) edge in $C \cap P_i$ and $q := C \setminus P_i$. For our proof it is sufficient to consider the subpath q of P_i' and therefore we refer to q as a *(tight) alternative* for e_α, defined by β and γ ($\beta \leq \gamma$; omitting P_i').

(N4) If q is a (tight) alternative for player i, defined by β and γ, we will denote the edges $\{e_\beta, \ldots, e_\gamma\} \cap P_i$ as the *edges which are substituted by q*. We call the nodes of these edges (except for the two endnodes of q) the *nodes which are substituted by q*.

(N5) We call a (tight) alternative q, defined by β and γ, a *left alternative*, if $\beta < \ell_1 + \ell_2$, and a *right alternative*, if $\gamma > \ell_1 + \ell_2 + m$.
If q is a left and a right alternative, we call q *large*; otherwise *small*.

(N6) If q (defined by β and γ) and p (defined by β' and γ') are two left alternatives of a player, we say that q is *smaller (larger)* than p, if $\gamma < \gamma'$ ($\gamma > \gamma'$). If q and p are right alternatives, q is *smaller (larger)* than p, if $\beta > \beta'$ ($\beta < \beta'$).

Note that if p is a smaller alternative than q, this means that p substitutes less commonly used edges than q. For illustration, consider the graph in Figure 3.12, where the solid lines form an optimal Steiner forest and the dashed lines are all existing tight alternatives. The alternatives q_1 (defined by β and γ), q_1' (defined by β' and γ') and $\bar{q}_1$ (defined by $\bar{\beta}$ and $\bar{\gamma}$) are all left alternatives of Player 1, where q_1 is large and $q_1', \bar{q}_1$ are small. Furthermore, q_1 is the largest left and q_1' is the smallest left alternative. We will sometimes need the smallest (or largest) alternative that substitutes a certain edge. For example, $\bar{q}_1$ is the smallest left alternative that substitutes e_α. For Player 2, q_2 (defined by μ and ν) and q_2' (defined by μ' and ν') are right alternatives where q_2' is larger than q_2.

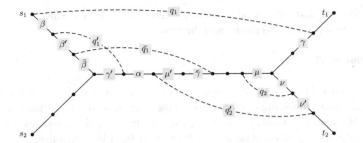

Fig. 3.12 Illustration for the used notation

3.3.5.1 Case *L*

In this subsection, we analyze Case L (that is, e is contained in the left part of F; ord$(e) \leq \ell_1 + \ell_2$) and conclude that this case cannot occur. Assume that Case L holds. We describe the subcase ord$(e) \leq \ell_1$ (i.e., $e \in P_1$) in detail, the other subcase $\ell_1 + 1 \leq \text{ord}(e) \leq \ell_1 + \ell_2$ ($e \in P_2$) follows analogously.

Since $e \notin P_2$ and e is not completely paid, we get $\sum_{i \in N_e(P)} \xi_{i,e} = \xi_{1,e} < c_e$. We now argue that Player 1 has a tight left alternative that substitutes e: If there is no such alternative, we can increase the cost share $\xi_{1,e}$ of Player 1 on edge e by some suitably small $\varepsilon > 0$ without changing the cost shares on the other edges and still get a feasible solution for LP(P). Since this solution has a higher objective function value, we get a contradiction to the optimality of the given cost shares for LP(P).

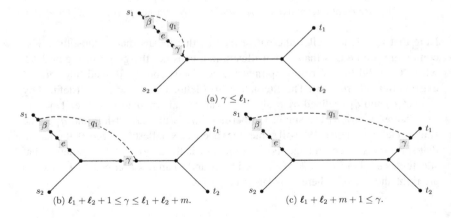

Fig. 3.13 Situation in Case L

Thus we can now assume that q_1, defined by β and γ, is a smallest left alternative (cf. Notation (N6)) of Player 1 for e. Depending on the location of e_γ, the situation looks as illustrated in Figure 3.13.

Regarding the cost shares of Player 1, we get (explanations below)

$$\xi_{1,e_\alpha} \begin{cases} = c_{e_\alpha}, & \beta \leq \alpha < \mathrm{ord}(e), \\ < c_e, & \alpha = \mathrm{ord}(e), \\ = 0, & \mathrm{ord}(e) < \alpha \leq \gamma, e_\alpha \in P_1. \end{cases}$$

The first two cases are clear due to the facts that the corresponding edges are not contained in P_2 and that e is the first edge that is not completely paid. The third case holds since q_1 is a *smallest* left alternative for e and the cost shares are *pushed to the left*: Assume there is an edge $e_\alpha \in P_1$ with $\mathrm{ord}(e) < \alpha \leq \gamma$ and $\xi_{1,e_\alpha} > 0$. Since the cost shares are PTL, there is no $\varepsilon > 0$ such that the operation $\mathrm{PUSHLEFT}(1, e, e_\alpha, \varepsilon)$ is feasible. This is only possible if there is a tight alternative for e which does not substitute e_α. But this alternative is then smaller than q_1; contradiction.

Now we construct a *cheaper* Steiner forest F^* than F, i.e., with strictly smaller cost than F, which is a contradiction to the optimality of F. To this end, consider the set of edges E^* resulting from F if the edges of q_1 are added, whereas the edges of $P_1 \setminus P_2$ which are substituted by q_1 are deleted. Figure 3.14 illustrates this for the case $\ell_1 + \ell_2 + 1 \leq \gamma \leq \ell_1 + \ell_2 + m$, where the deleted edges $\{e_\beta, \ldots, e, \ldots, e_{\ell_1}\}$ are dashed. Note that in general, E^* is not a Steiner forest, since q_1 may contain nodes or edges of $P_2 \setminus P_1$, which possibly yields a cycle in the subgraph induced by E^*. But of course, we obtain a Steiner forest by deleting suitable edges from E^*. Thus let $F^* \subseteq E^*$ be a Steiner forest. It remains to show that $C(F^*) < C(F)$. Considering the difference in costs, we get the desired inequality as follows (further explanations below):

$$
\begin{aligned}
c(F^*) - c(F) &\leq \sum_{f \in q_1} c_f - \sum_{e_\alpha \in P_1 \setminus P_2,\, \beta \leq \alpha \leq \gamma} c_{e_\alpha} \\
&= \sum_{e_\alpha \in P_1,\, \beta \leq \alpha \leq \gamma} \xi_{1,e_\alpha} - \sum_{e_\alpha \in P_1 \setminus P_2,\, \beta \leq \alpha \leq \gamma} c_{e_\alpha} \\
&= \sum_{\beta \leq \alpha \leq \mathrm{ord}(e)-1} \xi_{1,e_\alpha} + \xi_{1,e} + \sum_{e_\alpha \in P_1,\, \mathrm{ord}(e)+1 \leq \alpha \leq \gamma} \xi_{1,e_\alpha} - \sum_{e_\alpha \in P_1 \setminus P_2,\, \beta \leq \alpha \leq \gamma} c_{e_\alpha} \\
&< \sum_{\beta \leq \alpha \leq \mathrm{ord}(e)-1} c_{e_\alpha} + c_e - \sum_{e_\alpha \in P_1 \setminus P_2,\, \beta \leq \alpha \leq \gamma} c_{e_\alpha} \\
&\leq 0.
\end{aligned}
$$

The first inequality follows from the definition of F^* (respectively E^*) and the following equality is due to the fact that q_1 is a tight alternative. The strict inequality follows from our observations about the cost shares of Player 1, and the last inequality arises from the fact that at least the edges $\{e_\beta, \ldots, e_{\mathrm{ord}(e)-1}, e\}$ are deleted.

Overall we showed that Case L cannot occur.

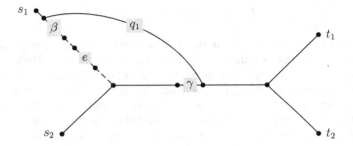

Fig. 3.14 Case L: Cheaper Steiner forest for $\ell_1 + \ell_2 + 1 \leq \gamma \leq \ell_1 + \ell_2 + m$

3.3.5.2 Case M

In this subsection, we analyze Case M (that is, e is contained in the middle part of F; $\ell_1 + \ell_2 + 1 \leq \mathrm{ord}(e) \leq \ell_1 + \ell_2 + m$) and conclude that this case cannot occur.

Assume that Case M holds. Using that e is in the middle part of F and not completely paid, we get $\sum_{i \in N_e(P)} \xi_{i,e} = \xi_{1,e} + \xi_{2,e} < c_e$. Since the cost shares are optimal, increasing $\xi_{1,e}$ or $\xi_{2,e}$ by any positive amount does not yield a feasible $LP(P)$-solution. This shows that both players need to have a tight alternative for e. We now distinguish between different cases regarding the properties of these tight alternatives. In each of these cases, we are able to construct a cheaper Steiner forest than F, contradicting the optimality of F. This shows that Case M cannot occur.

First, assume that there is a player $i \in \{1, 2\}$ who has a tight alternative q for e, defined by β and γ, which is *neither a left nor a right alternative* (see Figure 3.15). Using that q is tight, $\xi_{i,e_\alpha} \leq c_{e_\alpha}$ for each edge $e_\alpha \in P_i$ and $\xi_{i,e} < c_e$, we get that adding q and deleting the edges $\{e_\beta, \ldots, e, \ldots, e_\gamma\}$ which are substituted by q yields a cheaper Steiner forest F^* than F; contradiction (note that as in Case L, we possibly need to delete further edges to achieve a Steiner forest):

$$C(F^*) - C(F) \leq \sum_{f \in q} c_f - \sum_{\alpha = \beta}^{\gamma} c_{e_\alpha} = \sum_{\alpha = \beta}^{\gamma} \xi_{i,e_\alpha} - \sum_{\alpha = \beta}^{\gamma} c_{e_\alpha} < 0.$$

Fig. 3.15 Case M: A tight alternative which is neither left nor right

Since we derived a contradiction for the case that there is a tight alternative for e which is neither a left alternative nor a right, we can assume for the rest of the analysis that each tight alternative for e is a left or a right alternative (or both).

Now consider the case that there are two tight alternatives for e, one of Player 1 and one of Player 2, such that both alternatives substitute the *same commonly used edges*. Depending on whether all commonly used edges are substituted or not, Figure 3.16 illustrates the different possibilities, where q_1, defined by β and γ, is Player 1's tight alternative for e, and q_2, defined by μ and ν, is the tight alternative of Player 2. By adding q_1 and q_2, and deleting all edges which are substituted by q_1 and/or q_2, we again constructed an edge set which contains a cheaper Steiner forest F^* than F; contradiction (explanations below):

$$
\begin{aligned}
C(F^*) - C(F) \\
\leq \sum_{f \in q_1} c_f + \sum_{f \in q_2} c_f - \sum_{e_\alpha \in P_1 \setminus P_2, \beta \leq \alpha \leq \gamma} c_{e_\alpha} - \sum_{e_\alpha \in P_2 \setminus P_1, \mu \leq \alpha \leq \nu} c_{e_\alpha} - \sum_{e_\alpha \in P_1 \cap P_2, \beta \leq \alpha \leq \gamma} c_{e_\alpha} \\
= \sum_{e_\alpha \in P_1, \beta \leq \alpha \leq \gamma} \xi_{1,e_\alpha} + \sum_{e_\alpha \in P_2, \mu \leq \alpha \leq \nu} \xi_{2,e_\alpha} - \sum_{e_\alpha \in P_1 \setminus P_2, \beta \leq \alpha \leq \gamma} c_{e_\alpha} - \sum_{e_\alpha \in P_2 \setminus P_1, \mu \leq \alpha \leq \nu} c_{e_\alpha} \\
- \sum_{e_\alpha \in P_1 \cap P_2, \beta \leq \alpha \leq \gamma} c_{e_\alpha} \\
= \sum_{e_\alpha \in P_1 \setminus P_2, \beta \leq \alpha \leq \gamma} \xi_{1,e_\alpha} + \sum_{e_\alpha \in P_2 \setminus P_1, \mu \leq \alpha \leq \nu} \xi_{2,e_\alpha} + \sum_{e_\alpha \in P_1 \cap P_2, \beta \leq \alpha \leq \gamma} (\xi_{1,e_\alpha} + \xi_{2,e_\alpha}) \\
- \sum_{e_\alpha \in P_1 \setminus P_2, \beta \leq \alpha \leq \gamma} c_{e_\alpha} - \sum_{e_\alpha \in P_2 \setminus P_1, \mu \leq \alpha \leq \nu} c_{e_\alpha} - \sum_{e_\alpha \in P_1 \cap P_2, \beta \leq \alpha \leq \gamma} c_{e_\alpha} \\
< 0
\end{aligned}
$$

Here, we used that q_1 and q_2 are tight, the cost shares are bounded by the edge costs, and e is not completely paid.

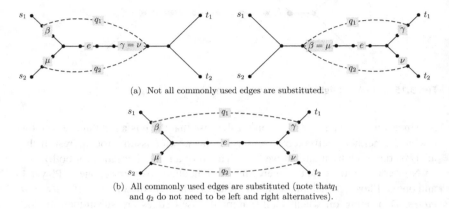

(a) Not all commonly used edges are substituted.

(b) All commonly used edges are substituted (note that q_1
and q_2 do not need to be left and right alternatives).

Fig. 3.16 Case M: Alternatives substituting the same commonly used edges

We derived a contradiction for the case that the players have tight alternatives substituting the same commonly used edges. Thus we can assume in the following that there are no tight alternatives of the players which substitute the same commonly used edges.

To complete our analysis of Case M, note that either both players have a tight *left* alternative for e, or there is a player having *only right* alternatives. We derive contradictions for both cases, which finally shows that Case M cannot occur.

First assume that both players have a tight left alternative for e. Let q_1 (defined by β and γ) be a smallest left alternative of Player 1 for e and q_2 (defined by μ and ν) a smallest left alternative of Player 2 for e (cf. Notation (N6)). Since we can assume that q_1 and q_2 do not substitute the same commonly used edges, one of these alternatives substitutes less commonly used edges than the other. We now describe the case that q_1 substitutes less edges commonly used edges than q_2 in detail (see Figure 3.17a), the case that q_2 substitutes less than q_1 can be treated analogously. Since q_1 and q_2 are tight alternatives, we get

$$\sum_{f \in q_1} c_f = \sum_{\alpha=\beta}^{\ell_1} \xi_{1,e_\alpha} + \sum_{\alpha=\ell_1+\ell_2+1}^{\gamma} \xi_{1,e_\alpha} \text{ and } \sum_{f \in q_2} c_f = \sum_{\alpha=\mu}^{\ell_1+\ell_2} \xi_{2,e_\alpha} + \sum_{e_\alpha \in P_2, \ell_1+\ell_2+1 \leq \alpha \leq \nu} \xi_{2,e_\alpha}.$$

Since e is the first edge which is not completely paid, since the cost shares are PTL, and since q_1 and q_2 are smallest left alternatives for e, we furthermore get

$$\xi_{1,e_\alpha} = \begin{cases} c_{e_\alpha}, & \beta \leq \alpha \leq \ell_1, \\ 0, & \text{ord}(e) < \alpha \leq \gamma, \end{cases} \qquad \xi_{2,e_\alpha} = \begin{cases} c_{e_\alpha}, & \mu \leq \alpha \leq \ell_1 + \ell_2, \\ 0, & \text{ord}(e) < \alpha \leq \nu, e_\alpha \in P_2 \end{cases}$$

and

$$\xi_{1,e_\alpha} + \xi_{2,e_\alpha} \begin{cases} = c_{e_\alpha}, & \ell_1 + \ell_2 + 1 \leq \alpha < \text{ord}(e), \\ < c_e, & \alpha = \text{ord}(e). \end{cases}$$

This shows that

$$\sum_{f \in q_1} c_f + \sum_{f \in q_2} c_f < \sum_{\alpha=\beta}^{\ell_1} c_{e_\alpha} + \sum_{\alpha=\mu}^{\ell_1+\ell_2} c_{e_\alpha} + \sum_{\alpha=\ell_1+\ell_2+1}^{\text{ord}(e)-1} c_{e_\alpha} + c_e.$$

Therefore adding q_1 and q_2, and deleting $\{e_\beta, \ldots, e_{\ell_1}, e_\mu, \ldots, e_{\ell_1+\ell_2}, e_{\ell_1+\ell_2+1}, \ldots, e, \ldots, e_\gamma\}$ (see Figure 3.17b) leads to a cheaper Steiner forest than F (after possibly deleting further edges); contradiction.

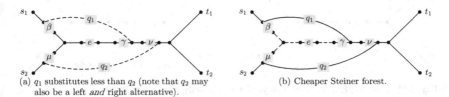

(a) q_1 substitutes less than q_2 (note that q_2 may   (b) Cheaper Steiner forest.
also be a left *and* right alternative).

Fig. 3.17 Case M: Both players have a left alternative

We derived a contradiction for the case that both players have a tight left alternative for e. It remains to consider the case that there is a player $i \in \{1, 2\}$ having *only* *right* tight alternatives for e. Let q, defined by β and γ, be a tight right alternative of player i for e which has the *smallest* possible value of γ among all such alternatives. The situation is illustrated in Figure 3.18a for the case that $i = 1$. Since q is a tight alternative, the sum of edge costs over edges contained in q equals the sum of cost shares of player i for the edges which are substituted by q. Regarding the cost shares of player i, we get (explanations below)

$$\xi_{i,e_\alpha} \begin{cases} \leq c_{e_\alpha}, & \beta \leq \alpha \leq \mathrm{ord}(e) - 1, \\ < c_{e_\alpha}, & \alpha = \mathrm{ord}(e), \\ = 0, & \mathrm{ord}(e) + 1 \leq \alpha \leq \gamma, e_\alpha \in P_i. \end{cases}$$

The first two cases are clear since the cost shares are bounded by the edge costs and e is not completely paid. The last case is due to the facts that there are no tight left alternatives for e, that we chose q with the smallest possible value of γ, and that the cost shares are PTL. Altogether we get that

$$\sum_{f \in q_1} c_f < \sum_{\alpha=\beta}^{\ell_1+\ell_2+m} c_{e_\alpha}$$

holds, showing that the edge set obtained from F by adding q and deleting the commonly used edges which are substituted by q, i.e., $\{e_\beta, \dots, e, \dots, e_{\ell_1+\ell_2+m}\}$ (see Figure 3.18b for the case that $i = 1$), contains a cheaper Steiner forest than F; contradiction.

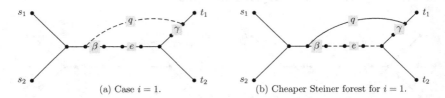

(a) Case $i = 1$.   (b) Cheaper Steiner forest for $i = 1$.

Fig. 3.18 Case M: Existence of a player i who has only right alternatives

Altogether, we showed that Case M cannot occur.

3.3.5.3 Case R

In this subsection, we analyze Case R (that is, e is contained in the right part of F; $\ell_1 + \ell_2 + m + 1 \leq \mathrm{ord}(e)$) and conclude that this case cannot occur. Assume that Case R holds. We describe the subcase $\mathrm{ord}(e) \leq \ell_1 + \ell_2 + m + r_1$ (i.e., $e \in P_1$) in detail, the other subcase $\ell_1 + \ell_2 + m + r_1 + 1 \leq \mathrm{ord}(e)$ ($e \in P_2$) follows analogously.

As already mentioned, the analysis of Case R is much more complicated than Case L and M. In this context, it is worth noting that we showed that the Cases L and M cannot occur without using that $(G, (s_1, t_1), (s_2, t_2))$ does not contain a BC as a subgraph. In contrast, during the analysis of Case R, several subcases occur where we need to analyze subgraphs which are "almost" Bad Configurations, so-called *Preliminary Bad Configurations (PBCs)* (see Definition 3.3.12 and also our explanation on page 40 in Section 3.3). Since $(G, (s_1, t_1), (s_2, t_2))$ does not contain a BC as a subgraph, these PBCs cannot be BCs. Nevertheless, the further analysis of the corresponding subcases is long and requires a lot of new ideas and technical definitions. Since the remaining case distinction of Case R is already quite long and complicated, we decided to analyze all subcases corresponding to PBCs in a separate subsection (Subsection 3.3.5.4) to make the analysis more readable. A further reason, why Case R is more complicated than the Cases L and M, is that we need additional properties for the cost shares to guarantee the existence of certain tight alternatives. We now introduce these properties, give a short motivation why they are useful, and show how to obtain cost shares with these properties. After this, we start with the formal analysis of Case R.

Definition 3.3.8 Let $(\xi_{i,f})_{i\in N, f\in P_i}$ be the given optimal solution for LP(P). For commonly used edges e_α and e_β, where $\alpha < \beta$, and $\varepsilon > 0$, the operation CHANGE$(\alpha, \beta, \varepsilon)$ increases the cost shares ξ_{1,e_α} and ξ_{2,e_β} by ε, and decreases ξ_{1,e_β} and ξ_{2,e_α} by ε (see Figure 3.19 for illustration). If these changes yield a feasible (and thus optimal) LP(P)-solution, we call CHANGE$(\alpha, \beta, \varepsilon)$ feasible (for $(\xi_{i,f})_{i\in N, f\in P_i}$). If there is an $\varepsilon > 0$ such that CHANGE$(\alpha, \beta, \varepsilon)$ is feasible, we say that CHANGE(α, β) is *feasible*, otherwise *infeasible* (for $(\xi_{i,f})_{i\in N, f\in P_i}$).

We call $(\xi_{i,f})_{i\in N, f\in P_i}$ *maximized for Player 2*, if the following two properties hold:

(2M) The sum of cost shares of Player 2 over all commonly used edges, i.e., $\sum_{f\in P_1\cap P_2} \xi_{2,f}$, is maximal among all optimal LP(P)-solutions which can be obtained from $(\xi_{i,f})_{i\in N, f\in P_i}$ by changing the cost shares of the commonly used edges.

(NC) For every pair of commonly used edges e_α and e_β with $\alpha < \beta$, the operation CHANGE(α, β) is infeasible.

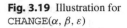

Fig. 3.19 Illustration for
CHANGE(α, β, ε)

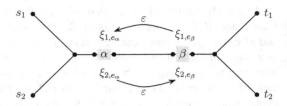

The following observations about cost shares which are maximized for Player 2 are
simple, but useful in the analysis of Case R.

Observation 3.3.9 Property (2M) implies that for each commonly used edge e_α
that Player 2 does not pay completely, i.e., with $\xi_{2,e_\alpha} < c_{e_\alpha}$, Player 2 has a tight
alternative for e_α: If there is no such alternative, there is an $\varepsilon > 0$ such that increasing
ξ_{2,e_α} and decreasing ξ_{1,e_α} by ε yields an optimal LP(P)-solution. Since this solution
is obtained from $(\xi_{i,f})_{i \in N, f \in P_i}$ by changing the cost shares for the commonly used
edge e_α, and the sum of cost shares of Player 2 over the commonly used edges is
larger than before, we get a contradiction to property (2M).

Property (NC) implies that for commonly used edges e_α and e_β with $\alpha < \beta$, and
with $\xi_{2,e_\alpha} > 0$ and $\xi_{1,e_\beta} > 0$, at least one of the following two tight alternatives
need to exist: A tight left alternative of Player 1 which substitutes e_α, but not e_β,
or a tight right alternative of Player 2 which substitutes e_β, but not e_α. To see this,
first note that any tight alternative of Player 1 which substitutes e_α, and any tight
alternative of Player 2 which substitutes e_β, needs to be a left or a right alternative
(or both). Otherwise, we get a contradiction to the optimality of F (as in a subcase
of Case M on page 54). Taking this into account, our statement above follows from
the fact that CHANGE(α, β) is feasible if and only if neither of the described tight
alternatives exists.

We now give some motivation why it is reasonable to consider cost shares which
are maximized for Player 2. Recall that the given cost shares $(\xi_{i,f})_{i \in N, f \in P_i}$ are an
optimal solution for LP(P). Therefore, it is not possible to change the cost shares
such that a feasible solution of LP(P) with higher sum of all cost shares is obtained.
In particular, we cannot change the cost shares for the commonly used edges such
that the optimality for LP(P) is preserved and there is no tight alternative of Player 1
for e (if this is possible, then we can additionally increase $\xi_{1,e}$ by some suitably

small positive amount and get a feasible solution with higher objective than the given optimal cost shares; contradiction). We now describe how the properties (2M) and (NC) are correlated with this argumentation. To this end, assume that q is the only alternative of Player 1 for e which is tight with respect to the given cost shares. We can assume w.l.o.g. that q also substitutes some commonly used edges (otherwise adding q and deleting all edges which are substituted by q leads to a cheaper Steiner forest than F; contradiction). Now assume that there is a commonly used edge e_α which is substituted by q, is not completely paid by Player 2, and Player 2 does not have a tight alternative for e_α. In particular this implies that property (2M) is not fulfilled (see Observation 3.3.9). But then, there is an $\varepsilon > 0$ such that increasing ξ_{2,e_α} and decreasing ξ_{1,e_α} by ε yields an optimal LP(P)-solution such that Player 1 does not have a tight alternative for e anymore (q is not tight anymore; see Figure 3.20a). As described above, this contradicts the optimality of the given cost shares. According to property (NC), assume that there are two commonly used edges e_α and e_β such that q substitutes e_β, but not e_α (in particular, this implies $\alpha < \beta$), and CHANGE(α, β) feasible. But this again contradicts the optimality of the given cost shares, since performing CHANGE($\alpha, \beta, \varepsilon$) (for some small enough $\varepsilon > 0$) yields that Player 1 does not have a tight alternative for e anymore (see Figure 3.20b).

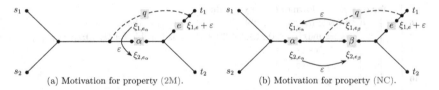

(a) Motivation for property (2M).   (b) Motivation for property (NC).

Fig. 3.20 Motivation for cost shares which are maximized for Player 2

In the following, we show that Algorithm MAX- 2 can be applied to obtain cost shares which are maximized for Player 2.

Algorithm 2: MAX- 2

    Input: Optimal solution $(\xi_{i,f})_{i\in\{1,2\},f\in P_i}$ for LP(P).

    Output: Transformed optimal solution $(\xi_{i,f})_{i\in\{1,2\},f\in P_i}$ for LP(P).

  1 Compute an optimal solution $(\xi'_{i,f})_{i\in\{1,2\},f\in P_1\cap P_2}$ for the linear program

$$\text{LP}'(P) \quad \max \sum_{f\in P_1\cap P_2} \xi'_{2,f}$$

$$\text{s.t.}: \xi'_{1,f} + \xi'_{2,f} = c_f \qquad\qquad\qquad \forall f \in P_1 \cap P_2$$

$$\sum_{f\in(P_1\cap P_2)\setminus P'_1} \xi'_{1,f} + \sum_{f\in(P_1\setminus P_2)\setminus P'_1} \xi_{1,f} \leq \sum_{f\in P'_1\setminus P_1} c_f \qquad \forall P'_1 \in \mathcal{P}_1$$

$$\sum_{f\in(P_1\cap P_2)\setminus P'_2} \xi'_{2,f} + \sum_{f\in(P_2\setminus P_1)\setminus P'_2} \xi_{2,f} \leq \sum_{f\in P'_2\setminus P_2} c_f \qquad \forall P'_2 \in \mathcal{P}_2$$

$$\xi'_{1,f} \geq 0, \xi'_{2,f} \geq 0 \qquad\qquad\qquad \forall f \in P_1 \cap P_2$$

    foreach $f \in P_1 \cap P_2$ **do**

  2 $\xi_{1,f} \leftarrow \xi'_{1,f}$;

  3 $\xi_{2,f} \leftarrow \xi'_{2,f}$;

  4 for $\beta = \ell_1 + \ell_2 + m$ down to $\ell_1 + \ell_2 + 1$ **do**

  5   **for** $\alpha = \beta - 1$ down to $\ell_1 + \ell_2 + 1$ **do**

  6   **if** CHANGE(α, β) is feasible for $(\xi_{i,f})_{i\in\{1,2\},f\in P_i}$ **then**

  7 $\varepsilon := \max\{\varepsilon' > 0 : \text{CHANGE}(\alpha, \beta, \varepsilon') \text{ is feasible for } (\xi_{i,f})_{i\in\{1,2\},f\in P_i}\}$;

  8 $\xi_{1,e_\alpha} \leftarrow \xi_{1,e_\alpha} + \varepsilon$;

  9 $\xi_{2,e_\beta} \leftarrow \xi_{2,e_\beta} + \varepsilon$;

 10 $\xi_{1,e_\beta} \leftarrow \xi_{1,e_\beta} - \varepsilon$;

 11 $\xi_{2,e_\alpha} \leftarrow \xi_{2,e_\alpha} - \varepsilon$;

Remark 3.3.10 It is clear that the constraints of the introduced linear program LP$'(P)$ exactly describe how the cost shares for the commonly used edges may be changed, while the optimality for LP(P) is preserved. Furthermore, LP$'(P)$ is feasible, since the given cost shares for the commonly used edges $(\xi_{i,f})_{i\in\{1,2\},f\in P_1\cap P_2}$ are a feasible solution. Since the objective function value of any feasible solution is bounded from above by the sum of edge costs over the commonly used edges, we conclude that an optimal solution for LP$'(P)$ exists.

To test if CHANGE(α, β) is feasible, and to compute ε, one can compute an optimal solution of a linear program which has one variable $\varepsilon \geq 0$ which is maximized, and the remaining constraints ensure that the cost shares resulting from $(\xi_{i,f})_{i\in\{1,2\},f\in P_i}$ if the cost shares for the edges e_α and e_β are changed as described in Line 8–11 of Algorithm MAX- 2, remain an optimal solution for LP(P).

Obviously, $\varepsilon = 0$ is a feasible solution for this problem, and ε is bounded by $\min\{c_{e_\alpha}, c_{e_\beta}\}$, thus an optimal solution exists. The optimal solution is zero if and only if CHANGE(α, β) is infeasible.

We now show that Algorithm MAX-2 actually computes cost shares which are maximized for Player 2.

Lemma 3.3.11 *Algorithm* MAX-2 *computes an optimal solution for LP(P) which is maximized for Player 2, that is, fulfills the properties (2M) and (NC).*

Proof First note that the computed cost shares are an optimal solution for LP(P). This follows from two facts: By construction of LP$'(P)$, replacing the cost shares for the commonly used edges by an optimal solution of LP$'(P)$ again yields an optimal solution for LP(P). Furthermore, the algorithm only performs feasible CHANGE-operations, and a CHANGE-operation does not influence the sum of all cost shares, i.e., the objective function of LP(P).

Now turn to the two properties (2M) and (NC). By construction of LP$'(P)$, property (2M) holds after the update of the cost shares for the commonly used edges in Line 2 and 3. Furthermore, any performed CHANGE-operation preserves property (2M), since the sum of cost shares of Player 2 over the commonly used edges is not changed. This shows that the computed cost shares fulfill property (2M).

It remains to show that property (NC) is fulfilled. Assume that this is not true and consider a pair of commonly used edges e_α and e_β with $\alpha < \beta$ such that CHANGE(α, β) is feasible for the computed cost shares $(\xi_{i,f})_{i\in\{1,2\}, f\in P_i}$. This means that the following four statements hold (see Observation 3.3.9): $\xi_{2,e_\alpha} > 0$, $\xi_{1,e_\beta} > 0$, there is no tight left alternative of Player 1 which substitutes e_α, but not e_β, and there is no tight right alternative of Player 2 which substitutes e_β, but not e_α. Let $(\bar{\xi}_{i,f})_{i\in N, f\in P_i}$ be the cost shares directly after we executed CHANGE$(\alpha, \beta, \varepsilon)$ during Algorithm MAX-2 (or detected infeasibility of CHANGE(α, β)). It is clear that $\bar{\xi}_{1,e_\beta} > 0$ holds since $\xi_{1,e_\beta} > 0$ and the cost share of Player 1 for e_β is never increased anymore during the remaining course of the algorithm. According to $\bar{\xi}_{2,e_\alpha}$, there are two cases: Either $\bar{\xi}_{2,e_\alpha} > 0$ or $\bar{\xi}_{2,e_\alpha} = 0$ holds. In the first case, i.e., $\bar{\xi}_{2,e_\alpha} > 0$, at least one of the following alternatives exists and is tight with respect to $(\bar{\xi}_{i,f})_{i\in N, f\in P_i}$ (see Observation 3.3.9): A left alternative of Player 1 which substitutes e_α, but not e_β, or a right alternative of Player 2 which substitutes e_β, but not e_α. But note that any such alternative is also tight with respect to $(\xi_{i,f})_{i\in N, f\in P_i}$, since any left alternative of Player 1, and any right alternative of Player 2, which is tight at some point in time during the course of the algorithm, obviously stays tight during the remaining course. This contradicts our assumption that CHANGE(α, β)

is feasible for the computed cost shares. Therefore $\overline{\xi}_{2,e_\alpha} = 0$ has to hold. Since $\xi_{2,e_\alpha} > 0$, there has to be a $\gamma < \alpha$ such that CHANGE(γ, α) is feasible during the course of Algorithm MAX- 2. Let $(\hat{\xi}_{i,f})_{i\in N, f\in P_i}$ be the cost shares directly after we executed CHANGE($\gamma, \beta, \varepsilon$) during Algorithm MAX- 2 (or detected infeasibility of CHANGE(γ, β)). Note that $\hat{\xi}_{2,e_\gamma} > 0$ holds since this cost share is not increased anymore until CHANGE($\gamma, \alpha, \varepsilon$) is performed, and $\hat{\xi}_{1,e_\beta} > 0$ holds since $\xi_{1,e_\beta} > 0$ and this cost share is not increased anymore during the remaining course of the algorithm. Therefore, at least one of the following alternatives exists and is tight with respect to $(\hat{\xi}_{i,f})_{i\in N, f\in P_i}$: A left alternative of Player 1 which substitutes e_γ, but not e_β, or a right alternative of Player 2 which substitutes e_β, but not e_γ. As already mentioned, such alternatives stay tight during the course of the algorithm. To complete the proof, we now consider both alternatives and derive a contradiction if the considered alternative exists. First assume that there is a left alternative of Player 1 which substitutes e_γ, but not e_β, and this alternative is tight with respect to $(\hat{\xi}_{i,f})_{i\in N, f\in P_i}$. This alternative also substitutes e_α since CHANGE(γ, α) is feasible during the execution of the algorithm. But this now contradicts our assumption that CHANGE(α, β) is feasible for the computed cost shares. Similarly, we derive a contradiction if there is a right alternative of Player 2 which substitutes e_β, but not e_γ, and which is tight with respect to $(\hat{\xi}_{i,f})_{i\in N, f\in P_i}$: This alternative does not substitute e_α, since CHANGE(γ, α) is feasible during the execution of the algorithm. But this again contradicts our assumption that CHANGE(α, β) is feasible for the computed cost shares. Altogether we derived a contradiction if property (NC) is not fulfilled for the computed cost shares, which completes the proof. $\square$

Analysis of Case R

We now start with the analysis of Case R. It turns out that we need to distinguish between a lot of different subcases, corresponding to different properties of certain tight alternatives. We derive contradictions for all subcases, showing that Case R cannot occur. Before we describe this in detail, we give a brief overview of the case distinction (cf. Figure 3.21, which contains the most important cases; not displayed cases easily yield contradictions (cheaper Steiner forests than F)). We consider two tight alternatives q_1 (for Player 1) and q_2 (for Player 2) with certain properties. Either q_2 is a left alternative ($R.3$), or a right one. If q_2 is a right alternative, we distinguish between the case that q_2 substitutes more ($R.2$) or less ($R.1$) commonly used edges than q_1. In $R.1$, we consider a third alternative q_2' which can be a left ($R.1.1$) or a right ($R.1.2$) alternative. In the subcases $R.1.2.1$ and $R.1.2.2$ we analyze why CHANGE(ρ, σ) (for certain edges e_ρ and e_σ) is not feasible: either Player 1 has a tight alternative which substitutes e_ρ, but not e_σ ($R.1.2.1$) or Player 2 has a tight alternative which substitutes e_σ, but not e_ρ ($R.1.2.2$). The subcases labelled with PBC1–PBC14 correspond to constructed subgraphs which are *Preliminary Bad Configurations* and these subcases are further analyzed in Subsection 3.3.5.4.

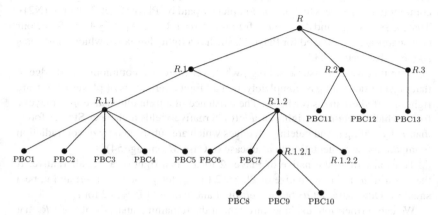

Fig. 3.21 Case R: Tree for case distinction (nodes are labelled with the (sub)cases)

We now analyze Case R, i.e., e is in the right part of F, in detail. Recall that we only need to analyze the case that $e \in P_1$, since the other case ($e \in P_2$) can be treated analogously. Assume w.l.o.g. that the given cost shares are maximized for Player 2 (otherwise, apply Algorithm MAX- 2). Since e is not completely paid, Player 1 needs to have a tight right alternative which substitutes e (if there is no such alternative, then increasing $\xi_{1,e}$ by some suitably small positive amount, while all other cost shares remain unchanged, yields a feasible solution for LP(P) with higher objective than the given optimal cost shares; contradiction). Let q_1, defined by β and γ, be a smallest such alternative (cf. Notation (N6)). It is clear that $\beta \leq \ell_1 + \ell_2 + m$ holds, i.e., q_1 also substitutes some commonly used edges: Otherwise, the edge set resulting from F by adding q_1 and deleting all edges which are substituted by q_1 has strictly smaller cost than F (since q_1 is tight and e is not completely paid), thus we get a cheaper Steiner forest than F after possibly deleting further "unnecessary" edges; contradiction.

In the remaining analysis of Case R, whenever we derive a contradiction by constructing a cheaper Steiner forest than F, we do not mention explicitly again the deletion of "unnecessary" edges. E.g., for the situation above, we simply say that we get a cheaper Steiner forest than F by adding q_1 and deleting the edges which are substituted by q_1.

If $\xi_{2,f} = c_f$ holds for all commonly used edges f which are substituted by q_1, i.e., Player 2 pays all those edges completely, we get a cheaper Steiner forest by adding q_1 and deleting $\{e_{\ell_1+\ell_2+m+1}, \ldots, e, \ldots, e_\gamma\}$. Therefore let e_α be the largest commonly used edge which is not completely paid by Player 2 (cf. Notation (N2)). That is, $\xi_{2,e_\alpha} < c_{e_\alpha}$ and $\xi_{2,e_\kappa} = c_{e_\kappa}$ for all $\kappa \in \{\alpha + 1, \ldots, \ell_1 + \ell_2 + m\}$. Since our cost shares are maximized for Player 2, she has a tight alternative which substitutes e_α (see Observation 3.3.9).

Note that we can assume w.l.o.g., whenever there is a commonly used edge f that player i does not pay completely, that all tight alternatives of player i for f are right or left alternatives (or both): The existence of a tight alternative q of player i for f which is neither a right nor a left alternative yields a cheaper Steiner forest than F by adding q and deleting all edges which are substituted by q; contradiction (compare our argumentation in a subcase of Case M on page 54).

Thus, any tight alternative of Player 2 for e_α is a right or a left alternative. If there is a *right* tight alternative of Player 2 for e_α, let q_2, defined by μ and ν, be a smallest. Otherwise, let q_2 be any *left* tight alternative of Player 2 for e_α.

We can furthermore exclude, throughout the remaining analysis of Case R, that there are two tight alternatives, one for each player, such that Player 1's alternative substitutes e and also some commonly used edges, and the alternative of Player 2 substitutes *the same* commonly used edges than q_1: Adding two such alternatives and deleting all edges which are substituted by them, yields a cheaper Steiner forest than F (compare our argumentation in a subcase of Case M on page 55).

Therefore, the following three different subcases, which are illustrated in Figure 3.22, cover the remaining situation in Case R:

Subcase $R.1$ q_2 is a right alternative and substitutes less commonly used edges than q_1 ($\nu \geq \ell_1 + \ell_2 + m + r_1 + 1$ and $\beta + 1 \leq \mu \leq \alpha$).

Subcase $R.2$ q_2 is a right alternative and substitutes more commonly used edges than q_1 ($\nu \geq \ell_1 + \ell_2 + m + r_1 + 1$ and $\ell_1 + 1 \leq \mu \leq \beta - 1$).

Subcase $R.3$ q_2 is a left alternative and q_2 is small, cf. Notation (N5) ($\ell_1 + 1 \leq \mu \leq \ell_1 + \ell_2$ and $\nu \leq \ell_1 + \ell_2 + m$).

We now further analyze the three subcases $R.1$, $R.2$ and $R.3$ and derive contradictions in all three cases, finally showing that Case R cannot occur. Start with Subcase $R.1$.

Subcase $R.1$

Recall that in Subcase $R.1$, the edge e_α is the largest commonly used edge which Player 2 does not completely, the alternative q_2 is a smallest right alternative of

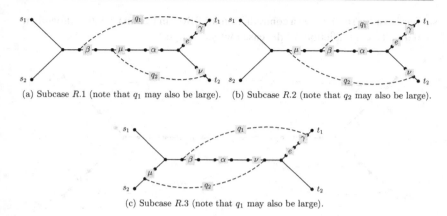

(a) Subcase $R.1$ (note that q_1 may also be large).   (b) Subcase $R.2$ (note that q_2 may also be large).

(c) Subcase $R.3$ (note that q_1 may also be large).

Fig. 3.22 Case R: Subcases $R.1$, $R.2$ and $R.3$

Player 2 for e_α, and q_2 substitutes less commonly used edges than q_1 (a smallest right alternative of Player 1 for e). If Player 2 pays all commonly used edges which are substituted by q_1, but not by q_2, we get a cheaper solution. Therefore let e_σ be the largest such edge which Player 2 does not pay completely. Since the cost shares are maximized by Player 2 there has to be a tight alternative for this edge. We now distinguish between the two cases that there is a tight left alternative for e_σ (Subcase $R.1.1$), or all tight alternatives for e_σ are right alternatives (Subcase $R.1.2$), and derive contradictions for both cases, showing that Subcase $R.1$ cannot occur. We start with Subcase $R.1.1$.

Subcase $R.1.1$
In Subcase $R.1.1$, there is a tight left alternative for e_σ. Let q_2', defined by μ' and ν', be a largest such one. We now distinguish between the two cases that q_1 is large or small, and derive contradictions for both cases, showing that Subcase $R.1.1$ cannot occur.

First assume that q_1 is large, i.e., $\beta \leq \ell_1$. If $\nu' \geq \alpha$ or $\nu' \leq \mu - 1$ we can use q_1, q_2' (and q_2 in the latter case) to get a cheaper solution. Therefore we can assume that $\nu' \in \{\mu, \ldots, \alpha - 1\}$. By the same argumentation, this has to hold for all tight left alternatives which substitute e_σ. Let q_2'' be a smallest such alternative. The situation is illustrated in Figure 3.23 (note that it is irrelevant for the argumentation whether $\mu' \leq \mu''$, as illustrated, or not). We denote this subcase as Subcase PBC1 (note that F, together with the alternatives q_1, q_2'' and q_2, build a subgraph which looks "very similar" to a BC) and derive a contradiction for PBC1 in Lemma 3.3.30. Altogether, we derived a contradiction for the case that q_1 is large.

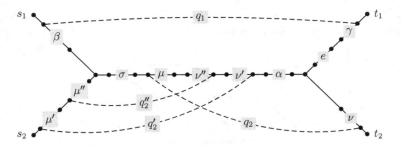

Fig. 3.23 Subcase PBC1

Now assume that q_1 is small. We first derive a contradiction if $v' \geq \mu$ holds. To this end, we need to further distinguish between the two cases that q_2' substitutes all commonly used edges (i.e., $v' \geq \ell_1 + \ell_2 + m$), or not ($\mu \leq v' < \ell_1 + \ell_2 + m$). We denote the latter case as Subcase PBC2 (see Figure 3.24; note that q_2' may also substitute e_α, but this is not relevant for the argumentation) and derive a contradiction for this case in Lemma 3.3.31.

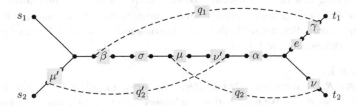

Fig. 3.24 Subcase PBC2

Now we consider the case that q_2' substitutes all commonly used edges ($v' \geq \ell_1 + \ell_2 + m$). If Player 1 pays the edges $e_{\ell_1 + \ell_2 + 1}, \ldots, e_{\beta - 1}$ completely or q_1 substitutes all commonly used edges, we get a cheaper Steiner forest by using q_1 and q_2'. Therefore let e_τ be the largest commonly used edge which is not substituted by q_1 and which Player 1 does not pay completely. Furthermore, let e_ρ be the smallest edge in $\{e_\beta, \ldots, e_{\mu - 1}\}$ which is not paid completely by Player 2. Thus we have two commonly used edges e_τ and e_ρ with $\tau < \rho$ and with $\xi_{1,e_\rho} > 0$ and $\xi_{2,e_\tau} > 0$. Since the cost shares are maximized for Player 2, CHANGE(τ, ρ) is not feasible. Thus at least one of the following two tight alternatives exists (see Observation 3.3.9): A tight left alternative for Player 1 which substitutes e_τ, but not e_ρ, or a tight right alternative for Player 2 which substitutes e_ρ, but not e_τ. We say that CHANGE(τ, ρ) *is (is not) feasible for Player 1* if there is no (a) tight alternative for Player 1 with the described properties. Analogous for Player 2: CHANGE(τ, ρ) *is (is not) feasible for Player 2* if there is no (a) tight alternative for Player 2 with the described properties. Using this, there is at least one player such that CHANGE(τ, ρ) is not feasible for this player. If CHANGE(τ, ρ) is not feasible for Player 2, we get a cheaper Steiner forest, and thus a contradiction, by using the tight right alternative for Player 2 which substitutes e_ρ, but not e_τ, together with q_1 (note that Player 1 completely pays the edges $e_{\tau+1}, \ldots, e_{\beta-1}$ and Player 2 completely pays $e_\beta, \ldots, e_{\rho-1}$). Thus we can assume that CHANGE(τ, ρ) is not feasible for Player 1. Let q_1', defined by β' and γ', be a smallest tight left alternative for Player 1 which substitutes e_τ, but not e_ρ. If $\tau \leq \gamma' \leq \beta - 1$, we can use q_1', q_1 and q_2' to construct a cheaper Steiner forest (since Player 1 completely pays the edges $e_{\tau+1}, \ldots, e_{\beta-1}$). Therefore, the situation is as illustrated in Figure 3.25 (note that $v' \in \{\ell_1 + \ell_2 + m\} \cup \{\ell_1 + \ell_2 + m + r_1 + 1, \ldots, \ell_1 + \ell_2 + m + r_1 + r_2\}$ may also be smaller than v, but this does not change the argumentation). We denote this case as Subcase PBC3 and derive a contradiction for this subcase in Lemma 3.3.32.

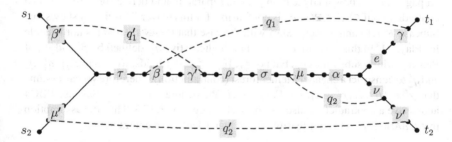

Fig. 3.25 Subcase PBC3

Overall we derived a contradiction if $v' \geq \mu$ holds (in the case that q_1 is small).

Recall that we are in Subcase $R.1.1$, i.e., q_2 (a smallest right alternative for e_α) substitutes less commonly used edges than q_1 (a smallest right alternative for e), and q_2' is a largest left alternative for e_σ. We already derived a contradiction for the case that q_1 is large, and for the case that q_1 is small *and* $v' \geq \mu$ holds. Thus we can now assume that q_1 is small and $\sigma \leq v' < \mu$ holds, see Figure 3.26. To complete the analysis of Subcase $R.1.1$, it remains to derive a contradiction for this case.

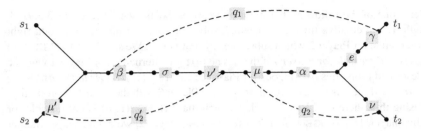

Fig. 3.26 Subcase $R.1.1$: Remaining situation

If q_1 substitutes all commonly used edges, or Player 1 completely pays all commonly used edges which are not substituted by q_1 (i.e., $e_{\ell_1+\ell_2+1}, \dots, e_{\beta-1}$), we get a contradiction, since using q_1, q_2' and q_2 yields a cheaper Steiner forest than F (note that Player 2 completely pays the edges $e_{\nu'+1}, \dots, e_{\mu-1}$). Thus we can now assume that e_ρ is the largest commonly used edge which is not substituted by q_1 and Player 1 does not pay completely. Thus we have two commonly used edges e_ρ and e_σ with $\rho < \sigma$ and with $\xi_{2,e_\rho} > 0$ and $\xi_{1,e_\sigma} > 0$. Since the cost shares are maximized for Player 2, CHANGE(ρ, σ) is not feasible. Thus there is a player such that CHANGE(ρ, σ) is not feasible for this player (see the notation introduced on page 69). We now analyze both players separately and derive for both players a contradiction if CHANGE(ρ, σ) is not feasible for this player. This finally shows that Subcase $R.1.1$ cannot occur. Start with the case that CHANGE(ρ, σ) is not feasible for Player 1. In this case, there is a tight left alternative q_1', defined by β' and γ', of Player 1 which substitutes e_ρ, but not e_σ. If $\gamma' \le \beta - 1$ holds, we can use q_1, q_1', q_2 and q_2' to construct a cheaper Steiner forest (contradiction). Therefore we can assume that $\gamma' \in \{\beta, \dots, \sigma - 1\}$, see Figure 3.27. We denote this case as Subcase PBC4 and derive a contradiction also for this case in Lemma 3.3.33. Thus the assumption that CHANGE(ρ, σ) is not feasible for Player 1 leads to a contradiction.

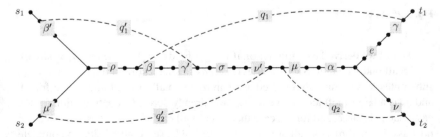

Fig. 3.27 Subcase PBC4

To complete the analysis of Subcase $R.1.1$, it remains to show that we also get a contradiction if CHANGE(ρ, σ) is not feasible for Player 2. Consider a tight right alternative $\bar{q}_2$, defined by $\bar{\mu}$ and $\bar{\nu}$, for Player 2 which substitutes e_σ, but not e_ρ. If $\bar{\mu} \leq \beta$ holds, we can use q_1 and $\bar{q}_2$ to get a cheaper Steiner forest than F, and thus a contradiction (note that Player 1 completely pays the edges $e_{\rho+1}, \ldots, e_{\beta-1}$ by the choice of ρ). Therefore $\bar{\mu} \in \{\beta + 1, \ldots, \sigma\}$ holds, see Figure 3.28 (note that it is irrelevant for the argumentation whether $\nu < \bar{\nu}$, as illustrated, or not). This case is denoted as Subcase PBC5 and we derive a contradiction for this case in Lemma 3.3.34, completing the analysis of Subcase $R.1.1$.

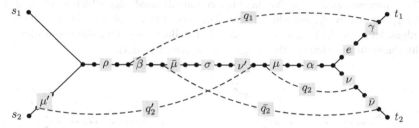

Fig. 3.28 Subcase PBC5

Subcase R.1.2
In Subcase $R.1.2$, there is no tight *left* alternative of Player 2 for e_σ, or, equivalently, all tight alternatives of Player 2 for e_σ are right alternatives. Recall that e_σ is the largest commonly used edge which is substituted by q_1, but not by q_2, and which Player 2 does not pay completely. The alternative q_1 is a smallest right alternative of Player 1 for e, and q_2 is a smallest right alternative of Player 2 for e_α, where e_α is the largest commonly used edge which Player 2 does not completely. The situation is illustrated in Figure 3.29.

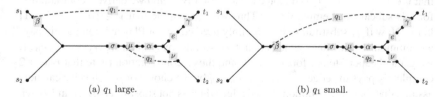

Fig. 3.29 Subcase R.1.2

Let q_2', defined by μ' and ν', be a largest right tight alternative of Player 2. Note that q_2' substitutes e_σ, since there is a tight right alternative of Player 2 for e_σ. Either

q_1 substitutes *more* commonly used edges than q_2', or *less* commonly used edges than q_2' (if both alternatives substitute the same commonly used edges, we get a cheaper Steiner forest than F, see the argumentation for a subcase of Case M on page 55). We derive a contradiction for both cases, which shows that Subcase $R.1.2$ cannot occur. Since we already showed that Subcase $R.1.1$ also cannot occur, this completes the analysis of Subcase $R.1$.

First consider the case that q_1 substitutes *more* commonly used edges than q_2'. If Player 2 completely pays all commonly used edges which are substituted by q_1, but not by q_2', we can use q_1 and q_2' to get a cheaper Steiner forest than F; contradiction. Therefore we can assume that there is a commonly used edge which is substituted by q_1, but not by q_2', and which Player 2 does not pay completely. Among all such edges, let e_ρ be the largest (see Figure 3.30 for illustration; note that it is irrelevant for the argumentation whether $\nu < \nu'$, as illustrated, or not).

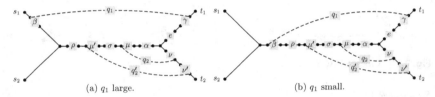

(a) q_1 large.   (b) q_1 small.

Fig. 3.30 Subcase R.1.2: q_1 substitutes more than q_2'

Since the cost shares are maximized for Player 2, there is a tight alternative of Player 2 for e_ρ. Furthermore, any such alternative is a *left* alternative, since q_2' is a *largest right* tight alternative of Player 2. Let $\bar{q}_2$, defined by $\bar{\mu}$ and $\bar{\nu}$, be a smallest left tight alternative of Player 2 for e_ρ. According to $\bar{\nu}$, we get that $\bar{\nu} \in \{\rho, \ldots, \sigma-1\}$ has to hold (since there is no left tight alternative of Player 2 for e_σ). The subcase that $\bar{\nu} \in \{\mu', \ldots, \sigma - 1\}$ is denoted as Subcase PBC6 and we derive a contradiction for this subcase in Lemma 3.3.35. Thus we can assume that $\bar{\nu} \in \{\rho, \ldots, \mu' - 1\}$ holds. Now if q_1 substitutes all commonly used edges, or Player 1 completely pays all commonly used edges which are not substituted by q_1, i.e., $e_{\ell_1+\ell_2+1}, \ldots, e_{\beta-1}$, we get a cheaper Steiner forest than F and thus a contradiction (note that Player 2 completely pays the edges $\{e_{\rho+1}, \ldots, e_{\mu'-1}\}$ by the choice of ρ). Thus we can now assume that there is a commonly used edge which is not substituted by q_1 and which Player 1 does not pay completely. Let e_ω be the largest such edge (see Figure 3.31 for illustration).

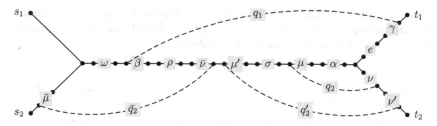

Fig. 3.31 Subcase $R.1.2$: q_1 substitutes more than q_2' (continued)

Now consider the two commonly used edges e_ω and e_ρ with $\xi_{2,e_\omega} > 0$ and $\xi_{1,e_\rho} > 0$. Since the cost shares are maximized for Player 2, there is at least one player such that CHANGE(ω, ρ) is not feasible for this player. For Player 2, there is no tight right alternative which substitutes e_ρ, but not e_ω, since any such alternative would be larger than q_2'. Therefore, we can assume that CHANGE(ω, ρ) is not feasible for Player 1. Let q_1', defined by β' and γ', be a tight left alternative of Player 1 which substitutes e_ω, but not e_ρ. According to γ', this means that $\gamma' \in \{\omega, \dots, \rho - 1\}$ holds. We now partition this set into two subsets and derive, for each subset, a contradiction if γ' is contained in this subset. If $\gamma' \in \{\omega, \dots, \beta - 1\}$ holds, using $q_1', q_1, \bar{q}_2$ and q_2' yields a cheaper Steiner forest than F; contradiction (note that Player 1 completely pays the edges $\{e_{\omega+1}, \dots, e_{\beta-1}\}$ by the choice of ω, and Player 2 completely pays the edges $\{e_{\rho+1}, \dots, e_{\mu'-1}\}$ by the choice of ρ). The other case, i.e., $\gamma' \in \{\beta, \dots, \rho - 1\}$, is denoted as Subcase PBC7 and we derive a contradiction for this case in Lemma 3.3.36. This way, we derived a contradiction for the subcase (of Subcase $R.1.2$), that q_1 substitutes *more* commonly used edges than q_2'.

To complete the analysis of Subcase $R.1.2$ (and also of Subcase $R.1$), it remains to derive a contradiction for the case that q_1 substitutes *less* commonly used edges than q_2'. If Player 1 completely pays all commonly used edges which are substituted by q_2', but not by q_1, we can use q_1 and q_2' to get a cheaper Steiner forest than F and thus a contradiction. We can thus assume that there is a commonly used edge which is substituted by q_2', but not by q_1, and which Player 1 does not pay completely. Let e_ρ be the largest such edge. Figure 3.32 illustrates the situation (note that it is irrelevant for the argumentation whether $\nu < \nu'$, as illustrated, or not). Recall that q_2' is small since there is no tight left alternative of Player 2 for e_σ. Consider the two commonly used edges e_ρ and e_σ with $\xi_{2,e_\rho} > 0$ and $\xi_{1,e_\sigma} > 0$. Since the cost shares are maximized for Player 2, there is at least one player such that CHANGE(ρ, σ) is not feasible for this player. Either CHANGE(ρ, σ) is feasible for Player 1, or not. In the former case, we get that CHANGE(ρ, σ) is *not* feasible for Player 2. We now distinguish between the following two cases: CHANGE(ρ, σ) is not feasible for

Player 1 (Subcase $R.1.2.1$), or CHANGE(ρ, σ) is feasible for Player 1, but not for Player 2 (Subcase $R.1.2.2$). We derive a contradiction in each case, which completes the analysis of Subcase $R.1.2$ (and also of Subcase $R.1$).

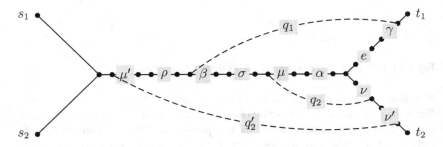

Fig. 3.32 Subcase $R.1.2$: q_1 substitutes less than q_2'

Subcase R.1.2.1
Besides the assumptions of Subcase $R.1.2$ described on page 71, we assume in Subcase $R.1.2.1$ that CHANGE(ρ, σ) is not feasible for Player 1. Here, e_ρ is the largest commonly used edge which is substituted by q_2' (a largest right tight alternative of Player 2), but not by q_1, and which Player 1 does not pay completely. Figure 3.32 illustrates the situation (note that the alternative q_2' is small, since there is no tight left alternative of Player 2 for e_σ). Since CHANGE(ρ, σ) is not feasible for Player 1, there is a tight left alternative of Player 1 which substitutes e_ρ, but not e_σ. Let q_1', defined by β' and γ', be a smallest such alternative. It is clear that $\gamma' \in \{\rho, \dots, \sigma - 1\}$ holds. We now argue that if q_2' substitutes all commonly used edges, or Player 2 completely pays the commonly used edges which are not substituted by q_2', i.e., $e_{\ell_1+\ell_2+1}, \dots, e_{\mu'-1}$, we get a cheaper Steiner forest than F, and thus a contradiction, by using q_1', q_1 and q_2': For the case that $\gamma' \in \{\rho, \dots, \beta - 1\}$, this follows from the fact that Player 1 completely pays the edges $e_{\rho+1}, \dots, e_{\beta-1}$ by the choice of ρ. For the other case according to γ', i.e., $\gamma' \in \{\beta, \dots, \sigma - 1\}$, we get the desired claim since the edges $e_\beta, \dots, e_{\gamma'}$ then all have cost zero (the case that there is such an edge with positive cost is denoted as Subcase PBC8 and we show in Lemma 3.3.37 that this case cannot occur). Thus we can now assume that there is a commonly used edge which is not substituted by q_2', and which Player 2 does not pay completely. Let e_ω be the largest such edge. Since the cost shares are maximized for Player 2, there is a tight alternative of Player 2 for e_ω. Furthermore, there is no such alternative which is a right alternative (since q_2' is a largest right tight alternative of Player 2). We can thus assume that $\bar{q}_2$, defined by $\bar{\mu}$ and $\bar{\nu}$, is a

smallest left tight alternative of Player 2 for e_ω. Since there is no tight left alternative of Player 2 for e_σ, we get that $\bar{\nu} \in \{\omega, \ldots, \sigma - 1\}$ holds. In Figure 3.33, we displayed the remaining situation, where we partitioned the possible locations for $\bar{\nu}$ into three subpaths. If $\bar{\nu}$ is contained in the thick black path, i.e., $\bar{\nu} \in \{\omega, \ldots, \mu' - 1\}$, we get a contradiction since using q_1', q_1, $\bar{q}_2$ and q_2' yields a cheaper Steiner forest than F (note that Player 2 completely pays the edges $e_{\omega+1}, \ldots, e_{\mu'-1}$ by the choice of ω). The case that $\bar{\nu}$ is contained in the dotted path, i.e., $\bar{\nu} \in \{\mu', \ldots, \rho - 1\}$, is denoted as Subcase PBC9 and we derive a contradiction for this case in Lemma 3.3.38. Thus we can assume for the remaining analysis that $\bar{\nu}$ is contained in the grey subpath, i.e., $\bar{\nu} \in \{\rho, \ldots, \sigma - 1\}$.

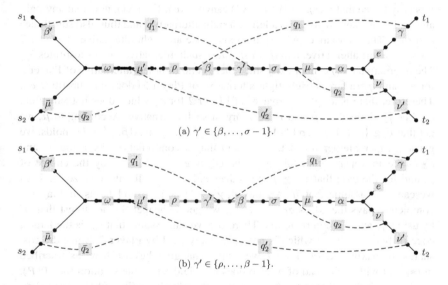

(a) $\gamma' \in \{\beta, \ldots, \sigma - 1\}$.

(b) $\gamma' \in \{\rho, \ldots, \beta - 1\}$.

Fig. 3.33 Subcase $R.1.2.1$: Remaining situations

Consider the three commonly used edges e_ω, e_ρ and e_σ, and note that $\xi_{1,e_\omega} > 0$, $\xi_{2,e_\rho} > 0$ and $\xi_{1,e_\sigma} > 0$ holds. Since the cost shares are maximized for Player 2, it is not possible to change the cost shares for the commonly used edges such that the resulting solution for LP(P) is optimal, and the sum of cost shares of Player 2 over the commonly used edges is larger than before (property (2M)). This implies that, for any $\varepsilon > 0$, the following changes of the cost shares do not yield a feasible solution for LP(P):

$$(*) \begin{cases} \text{Increase } \xi_{1,e_\rho} \text{ and decrease } \xi_{1,e_\omega} \text{ and } \xi_{1,e_\sigma} \text{ by } \varepsilon. \\ \text{Decrease } \xi_{2,e_\rho} \text{ and increase } \xi_{2,e_\omega} \text{ and } \xi_{2,e_\sigma} \text{ by } \varepsilon. \end{cases}$$

This implies that at least one of the following four alternatives (A1), (A2), (A3) and (A4) exists:

(A1) A tight alternative of Player 1 for e_ρ which neither substitutes e_ω, nor e_σ.
(A2) A tight alternative of Player 2 for e_ω which does not substitute e_ρ.
(A3) A tight alternative of Player 2 for e_σ which does not substitute e_ρ.
(A4) A tight alternative of Player 2 which substitutes e_ω and e_σ.

It is easy to see that (A1), (A2) and (A4) cannot exist: For (A1), note that any tight alternative of Player 1 for e_ρ is a left or a right alternative, and thus also substitutes e_ω or e_σ. The alternative (A2) cannot exist since any tight alternative of Player 2 for e_ω is a left alternative, and $\bar{q}_2$ (a *smallest* such alternative) also substitutes e_ρ. The alternative (A4) cannot exist since there is no tight left alternative of Player 2 for e_σ, and q_2' (a largest right tight alternative of Player 2) does not substitute e_ω. Therefore, there is a tight alternative of Player 2 for e_σ which does not substitute e_ρ. Let $\widehat{q}_2$, defined by $\widehat{\mu}$ and $\widehat{\nu}$, be a largest such alternative. According to $\widehat{\mu}$, we get that $\widehat{\mu} \in \{\rho + 1, \ldots, \sigma\}$ holds. If $\widehat{\mu} \in \{\rho + 1, \ldots, \max\{\beta, \gamma' + 1\}\}$ holds, we get a cheaper Steiner forest than F, and thus a contradiction, by using q_1 and $\widehat{q}_2$ (since Player 1 completely pays the edges $e_{\rho+1}, \ldots, e_{\beta-1}$ by the choice of ρ, and for the case that $\gamma' \geq \beta$, the edges $e_\beta, \ldots, e_{\gamma'}$ all have cost zero). Thus we can now assume that $\widehat{\mu} \in \{\max\{\beta, \gamma' + 1\} + 1, \ldots, \sigma\}$ holds. If Player 2 completely pays the edges $e_\beta, \ldots, e_{\widehat{\mu}-1}$, we get a cheaper Steiner forest than F by using q_1 and $\widehat{q}_2$; contradiction. Therefore we can assume that $e_{\sigma'}$ is the largest edge in $\{e_\beta, \ldots, e_{\widehat{\mu}-1}\}$ which is not completely paid by Player 2. Since the cost shares are maximized for Player 2, we get that changing the cost shares as described in $(*)$, but with σ' instead of σ, also does not yield a feasible solution for LP(P). We again conclude that at least one of the four tight alternatives (A1)–(A4) exists. The existence of (A1) and (A2) can be excluded by the same argumentation as before. Furthermore, there cannot be an alternative (A3), since any such alternative also substitutes e_σ, and is larger than $\widehat{q}_2$, a *largest* tight alternative of Player 2 for e_σ which does not substitute e_ρ. Therefore, there is a tight left alternative of Player 2 for $e_{\sigma'}$ (note that Player 2 does not have a tight right alternative which substitutes e_ω). Let $\widetilde{q}_2$, defined by $\widetilde{\mu}$ and $\widetilde{\nu}$, be any such alternative. Note that $\widetilde{q}_2$ does not substitute e_σ, since there is no tight left alternative of Player 2 for e_σ. Therefore, $\widetilde{\nu} \in \{\sigma', \ldots, \sigma - 1\}$ holds. We now subdivide this set into two subsets and derive, for each subset, a contradiction if $\widetilde{\nu}$ is contained in this subset, showing that Subcase $R.1.2.1$ cannot occur. If $\tilde{v} \in \{\sigma', \ldots, \widehat{\mu}-1\}$, we can construct a cheaper Steiner Forest than F by using q'_1, q_1, $\tilde{q}_2$ and $\widehat{q}_2$. The other subcase according to $\tilde{v}$, i.e., $\tilde{v} \in \{\widehat{\mu}, \ldots, \sigma - 1\}$, is denoted as Subcase PBC10 and we derive a contradiction for this case in Lemma 3.3.39. This completes our analysis of Subcase $R.1.2.1$.

Subcase R.1.2.2
Besides the assumptions of Subcase $R.1.2$ described on page 71, we assume in Subcase $R.1.2.2$ that CHANGE(ρ, σ) is feasible for Player 1, but not for Player 2. Here, e_ρ is the largest commonly used edge which is substituted by q'_2 (a largest right tight alternative of Player 2), but not by q_1, and which Player 1 does not pay completely. Figure 3.32 illustrates the situation (note that the alternative q'_2 is small, since there is no tight left alternative of Player 2 for e_σ). Since CHANGE(ρ, σ) is not feasible for Player 2, there is a tight right alternative for Player 2 which substitutes e_σ, but not e_ρ. Among all such alternatives, let $\bar{q}_2$, defined by $\bar{\mu}$ and $\bar{v}$, be a largest. According to $\bar{\mu}$, we get that $\bar{\mu} \in \{\rho + 1, \ldots, \sigma\}$ holds. If $\bar{\mu} \in \{\rho + 1, \ldots, \beta\}$, we get a contradiction since using q_1 and $\bar{q}_2$ yields a cheaper Steiner forest than F (note that Player 1 completely pays the edges $\{e_{\rho+1}, \ldots, e_{\beta-1}\}$ by the choice of ρ). Thus we can now assume that $\bar{\mu} \in \{\beta + 1, \ldots, \sigma\}$ holds. If Player 2 completely pays all commonly used edges which are substituted by q_1, but not by $\bar{q}_2$, we get a cheaper Steiner forest than F, and thus a contradiction, by using q_1 and $\bar{q}_2$. Therefore we can assume that e_ω is a commonly used edge which is substituted by q_1, but not by $\bar{q}_2$, and which Player 2 does not pay completely. The situation is illustrated in Figure 3.34 (note that the ordering of v, v' and $\bar{v}$ is irrelevant for the argumentation, therefore we have chosen $v' = \bar{v}$ to simplify the illustration).

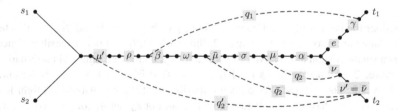

Fig. 3.34 Subcase R.1.2.2: Remaining situation

Note that $\xi_{2,e_\rho} > 0$ and $\xi_{1,e_\omega} > 0$ holds for the two commonly used edges e_ρ and e_ω. Furthermore, CHANGE(ρ, ω) is feasible for Player 1, since any tight left alternative of Player 1 which substitutes e_ρ, but not e_ω, in particular does not substitute e_σ, and such an alternative cannot exist because CHANGE(ρ, σ) is feasible for Player 1. Using that the cost shares are maximized for Player 2, we conclude that CHANGE(ρ, ω) is not feasible for Player 2. But this contradicts our choice of $\bar{q}_2$ as the largest right alternative of Player 2 which substitutes e_σ, but not e_ρ: Any tight right alternative of Player 2 which substitutes e_ω, but not e_ρ, also substitutes e_σ, and is larger than $\bar{q}_2$. This finally shows that Subcase $R.1$ cannot occur, and thus completes our analysis of this case.

Subcase $R.2$

Recall that in Subcase $R.2$, the edge e_α is the largest commonly used edge which Player 2 does not completely, the alternative q_2 is a smallest right alternative of Player 2 for e_α, and q_2 substitutes more commonly used edges than q_1 (a smallest right alternative of Player 1 for e). If Player 1 completely pays all edges of the commonly used path which are substituted by q_2, but not by q_1, then using q_1 and q_2 yields a cheaper Steiner forest than F; contradiction. Thus let e_σ be the largest commonly used edge which is substituted by q_2, but not by q_1, and which Player 1 does not pay completely. Figure 3.35 illustrates the situation.

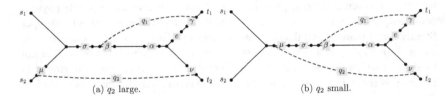

(a) q_2 large.   (b) q_2 small.

Fig. 3.35 Subcase R.2

Consider the commonly used edges e_σ and e_α with $\xi_{2,e_\sigma} > 0$ and $\xi_{1,e_\alpha} > 0$. The cost shares are maximized for Player 2, thus CHANGE(σ, α) is not feasible. Since q_2 is a *smallest* right alternative of Player 2 for e_α, there is no tight right alternative of Player 2 which substitutes e_α, but not e_σ. That is, CHANGE(σ, α) is feasible for Player 2. Thus CHANGE(σ, α) is not feasible for Player 1. Among all tight left alternatives for Player 1 which substitute e_σ, but not e_α, let q_1' (defined by β' and γ') be a smallest. According to γ', either $\beta \leq \gamma' \leq \alpha - 1$, or $\sigma \leq \gamma' \leq \beta - 1$ holds. We denote the first case as Subcase PBC11 (see Figure 3.36) and derive a contradiction for this case in Lemma 3.3.40. Therefore we can now assume that $\sigma \leq \gamma' \leq \beta - 1$ holds. Let q_2', defined by μ' and ν', be a *largest* right alternative of Player 2. If q_2' substitutes all commonly used edges, or Player 2 completely pays all commonly used edges which are not substituted by q_2', we get a cheaper Steiner forest than F, and thus a contradiction, by

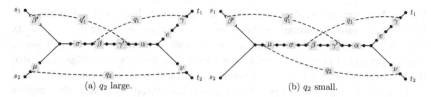

(a) q_2 large.   (b) q_2 small.

Fig. 3.36 Subcase PBC11

using q_1, q_1' and q_2' (note that Player 1 completely pays the edges $e_{\gamma'+1}, \ldots, e_{\beta-1}$). Therefore there is a commonly used edge which is not substituted by q_2', i.e., in $\{e_{\ell_1+\ell_2+1}, \ldots, e_{\mu'-1}\}$, which Player 2 does not pay completely. Let e_τ be the largest such edge. Since the cost shares are maximized for Player 2, there is a tight alternative of Player 2 for e_τ. Any such alternative is a *left* alternative, since q_2' is Player 2's largest right alternative (and e_τ is not substituted by q_2'). Let $\bar{q}_2$, defined by $\bar{\mu}$ and $\bar{\nu}$, be a smallest tight left alternative of Player 2 for e_τ. Note that $\bar{q}_2$ is no right alternative (q_2' is a *largest* right alternative of Player 2) and thus $\bar{\nu} \in \{\tau, \ldots, \ell_1 + \ell_2 + m\}$ holds. We now partition this set into different subsets and derive, for each subset, a contradiction if $\bar{\nu}$ is located in this subset. This completes the analysis of Subase $R.2$ since we showed that this case can not occur. To this end, consider Figure 3.37. If $\bar{\nu}$ is located in one of the thick paths, i.e., $\bar{\nu} \in \{\tau, \ldots, \mu' - 1\}$ or $\bar{\nu} \in \{\alpha, \ldots, \ell_1 + \ell_2 + m\}$, we get a contradiction since using q_1', q_1, q_2' (and $\bar{q}_2$ if $\bar{\nu}$ is in the first set) yields a cheaper Steiner forest than F (note that Player 2 completely pays the edges $e_{\tau+1}, \ldots, e_{\mu'-1}$ by the choice of τ, and $e_{\alpha+1}, \ldots, e_{\ell_1+\ell_2+m}$ by the choice of α). The case that $\bar{\nu}$ is located in the dotted path, i.e., $\bar{\nu} \in \{\mu', \ldots, \sigma - 1\}$, is denoted as Subcase PBC12 and we derive a contradiction for this case in Lemma 3.3.41. It remains to analyze the case that $\bar{\nu} \in \{\sigma, \ldots, \alpha - 1\}$ holds.

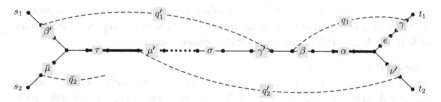

Fig. 3.37 Subcase $R.2$: Remaining situation

For this last case, we use the same argumentation as in Subcase $R.1.2.1$ on page 75:
Consider the three commonly used edges e_τ, e_σ and e_α, and note that $\xi_{1,e_\tau} > 0$,
$\xi_{2,e_\sigma} > 0$ and $\xi_{1,e_\alpha} > 0$ holds. Since the cost shares are maximized for Player 2,
it is not possible to change the cost shares for the commonly used edges such that
the resulting solution for LP(P) is optimal, and the sum of cost shares of Player 2
over the commonly used edges is larger than before (property (2M)). This implies
that, for any $\varepsilon > 0$, the following changes of the cost shares do not yield a feasible
solution for LP(P):

$$\text{Increase } \xi_{1,e_\sigma} \text{ and decrease } \xi_{1,e_\alpha} \text{ and } \xi_{1,e_\tau} \text{ by } \varepsilon.$$
$$\text{Decrease } \xi_{2,e_\sigma} \text{ and increase } \xi_{2,e_\alpha} \text{ and } \xi_{2,e_\tau} \text{ by } \varepsilon.$$

This implies that at least one of the following four alternatives exists:

A tight alternative of Player 1 for e_σ which neither substitutes e_τ, nor e_α.
A tight alternative of Player 2 for e_τ which does not substitute e_σ.
A tight alternative of Player 2 for e_α which does not substitute e_σ.
A tight alternative of Player 2 which substitutes e_α and e_τ.

It is clear that the first alternative cannot exist, since any tight alternative of Player 1
for e_σ is a left or a right alternative, and thus also substitutes e_τ or e_α. For the second
alternative, note that any tight alternative of Player 2 for e_τ is a left alternative, and
$\bar{q}_2$ (a *smallest* such alternative) also substitutes e_σ. Therefore, the second alternative
is also not possible. The third alternative also does not exist: Any such alternative is
a right alternative for e_α and is smaller than q_2, a smallest right tight alternative of
Player 2 for e_α. Finally, the fourth alternative cannot exist since there is no tight right
alternative of Player 2 for e_τ, and any left tight alternative of Player 2 for e_α leads to a
cheaper Steiner forest than F (together with q_1' and q_1; note that Player 1 completely
pays the edges $e_{\sigma+1}, \ldots, e_{\beta-1}$ by the choice of σ, and Player 2 completely pays the
edges $e_{\alpha+1}, \ldots, e_{\ell_1+\ell_2+m}$ by the choice of α). We thus derived a contradiction to
the fact that at least one of the four alternatives exists, which completes the analysis
of Subcase $R.2$ and shows that this subcase cannot occur.

Subcase $R.3$

Recall that in Subcase $R.3$, the edge e_α is the largest commonly used edge which
Player 2 does not completely, Player 2 does not have a tight right alternative which
substitutes e_α, and q_2 is a tight left alternative for Player 2 which substitutes e_α. If
q_1 substitutes all commonly used edges, or Player 1 completely pays all commonly
used edges which are not substituted by q_1, i.e., $e_{\ell_1+\ell_2+1}, \ldots, e_{\beta-1}$, we get a cheaper

Steiner forest than F, and thus a contradiction, by using q_1 and q_2 (note that Player 2 completely pays the edges $e_{\alpha+1}, \ldots, e_{\ell_1+\ell_2+m}$ by the choice of α). Thus we can assume that there is a commonly used edge which is not substituted by q_1 and which Player 1 does not pay completely. Let e_σ be the largest such edge. Figure 3.38 illustrates the situation.

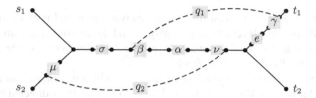

Fig. 3.38 Subcase R.3

Note that $\xi_{2,e_\sigma} > 0$ and $\xi_{1,e_\alpha} > 0$ holds for the two commonly used edges e_σ and e_α. Since the cost shares are maximized for Player 2, there has to be player such that CHANGE(σ, α) is not feasible for this player. For Player 2, CHANGE(σ, α) is feasible, since there is no tight *right* alternative of Player 2 for e_α (and thus in particular no tight right alternative which substitutes e_α, but not e_σ). Thus CHANGE(σ, α) is not feasible for Player 1. Among all tight left alternatives for Player 1 which substitute e_σ, but not e_α, let q_1', defined by β' and γ', be a smallest. According to γ', either $\sigma \le \gamma' \le \beta - 1$, or $\beta \le \gamma' \le \alpha - 1$ holds. We now show that both cases lead to a contradiction, which overall shows that Subcase $R.3$ cannot occur. In the first case, i.e., $\sigma \le \gamma' \le \beta - 1$, we get a cheaper Steiner forest than F by using q_1', q_1 and q_2 (note that Player 1 completely pays the edges $e_{\sigma+1}, \ldots, e_{\beta-1}$ by the choice of σ). The other case, i.e., $\beta \le \gamma' \le \alpha - 1$, is denoted as Subcase PBC13 and we derive a contradiction for this subcase in Lemma 3.3.42.

Overall we showed that neither of the subcases $R.1$, $R.2$ and $R.3$ can occur, which finally shows that Case R cannot occur. Since we already showed that the Cases L and M also cannot occur, this shows that the assumption, that $(G, (s_1, t_1), (s_2, t_2))$ is not strongly efficient, is false. This completes the whole proof for the direction $(1) \Rightarrow (2)$ of Theorem 3.3.1.

3.3.5.4 Analysis of Constructed PBCs

In this subsection, we derive contradictions for the subcases PBC1–PBC13 of Case R, showing that these subcases cannot occur. Note that in each of these subcases, we detected a subgraph of $(G, (s_1, t_1), (s_2, t_2))$ which is "almost" a Bad

Configuration (see also our explanation on page 40 in Section 3.3). We first give a formal definition of so-called *Preliminary Bad Configurations (PBCs)*.

Definition 3.3.12 (Preliminary Bad Configurations (PBCs)). Let P_u, P_ℓ, q_1, q_2 and q_3 be paths in $(G, (s_1, t_1), (s_2, t_2))$. If the following properties (1)–(7) are satisfied, we call $(P_u, P_\ell, q_1, q_2, q_3)$ a *Preliminary Bad Configuration (PBC)*:

(1) $P_u \cup P_\ell$ is a Steiner forest (that is, the endvertices u and v of P_u are source and sink of one player (the *upper* player) and the endvertices w and x of P_ℓ are source and sink of the other player (the *lower* player), and the subgraph induced by $P_u \cup P_\ell$ is cycle-free).

(2) There is at least one *commonly used edge*, i.e., which is contained in P_u and in P_ℓ, and if we consider P_u as directed from u to v and P_ℓ as directed from w to x, this induces the same direction on the commonly used edges.

(3) The subpaths R_u, L_ℓ and R_ℓ each contain at least one edge, where R_u is the subpath of P_u which connects the commonly used edges with v, and L_ℓ and R_ℓ are the subpaths of P_ℓ connecting w, respectively x, with the commonly used edges (see Figure 3.39a) for illustration.

(4) The path q_1 is an alternative of the upper player[9] which substitutes commonly used edges and edges from R_u. If q_1 additionally substitutes edges from L_u, we call q_1 *large*, otherwise *small* (see Figures 3.39b and 3.39c).

(5) The path q_2 is an alternative of the lower player which substitutes edges from L_ℓ and commonly used edges, but not all commonly used edges (see Figure 3.39d).

(6) The path q_3 is an alternative of the lower player which substitutes edges from R_ℓ and commonly used edges, but not all commonly used edges (see Figure 3.39e).

(7) There is at least one commonly used edge which is substituted by q_2 and q_3, and each such edge is also substituted by q_1. Furthermore, there is at least one commonly used edge which is substituted by q_1 and q_2, but not by q_3. This leads to the two PBCs displayed in Figures 3.39f and 3.39g.

Note that alternatives are internal node-disjoint with their respective paths. Therefore, q_1 is internal node-disjoint with P_u, and q_2 and q_3 are internal node-disjoint with P_ℓ. However, alternatives do not need to be internal node-disjoint *with the other player's path*: Nodes of P_ℓ which are only used by the lower player may be contained in q_1, and nodes of P_u which are only used by the upper player may be part of q_2 or q_3. Furthermore, the paths q_1, q_2 and q_3 do not have to be internal node-disjoint

---

[9]In the sense we used before, i.e., q_1 is internal node-disjoint with P_u, and the union with P_u contains a (unique) cycle.

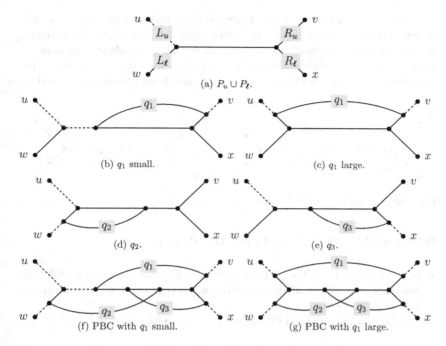

Fig. 3.39 Illustration for the definition of PBCs (solid lines always contain at least one edge, whereas dashed lines may consist of only one node)

with each other. These properties determine whether a PBC $(P_u, P_\ell, q_1, q_2, q_3)$ is a Bad Configuration, or not.

In each of the Subcases PBC1–PBC13 we detected a PBC. Since $(G, (s_1, t_1), (s_2, t_2))$ does not contain a BC as a subgraph, these PBCs cannot be BCs. Our proof strategy is as follows. We define twelve types of PBCs which are no BCs, so-called *No Bad Configurations (NBCs)* (see Definition 3.3.14) and show that each PBC has to be an NBC, if we exclude the existence of BCs (see Lemma 3.3.15). Then we derive properties for each type of NBC (see Lemma 3.3.17–Lemma 3.3.29) which are finally used to get contradictions to the properties of the Subcases PBC1–PBC13 (see Lemma 3.3.30–Lemma 3.3.42).

Bad Configurations

For a better understanding of the PBCs which are no BCs, it is useful to first give a definition of BCs in terms of properties of PBCs.

To this end, we introduce directed versions $\vec{q_1}$, $\vec{q_2}$ and $\vec{q_3}$ of the paths q_1, q_2 and q_3, where we choose the direction "from left to right" (cf. Figure 3.40). Furthermore, we sometimes need to subdivide the paths q_1, q_2 and q_3 in subpaths. In terms of notation, we will always use α_i for subpaths of q_1, β_i for subpaths of q_2 and γ_i for subpaths of q_3. The (directed) subpaths of $\vec{q_1}$, $\vec{q_2}$ and $\vec{q_3}$ are written as $\vec{\alpha_i}$, $\vec{\beta_i}$ and $\vec{\gamma_i}$.

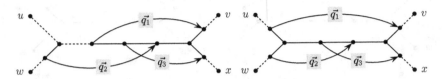

Fig. 3.40 Directed versions of the alternatives q_1, q_2 and q_3

Definition 3.3.13 (BCs). We call a PBC $(P_u, P_\ell, q_1, q_2, q_3)$ a

- *BCi*, for an $i \in \{1, 2, 3, 4\}$, if the properties of BCi (listed below) are fulfilled. Note that some of the BCis have different subtypes which are also described below.
- *Bad Configuration (BC)*, if $(P_u, P_\ell, q_1, q_2, q_3)$ is a BCi for an $i \in \{1, 2, 3, 4\}$.

For an illustration of all different BCs, see Figure 3.41.

BC1: (with subtypes a and b)

- q_1, q_2 and q_3 are pairwise internal node-disjoint.
- q_1, q_2 and q_3 are internal node-disjoint with $P_u \cup P_\ell$.
- BC1a: q_1 small; BC1b: q_1 large.

BC2: (with subtypes a, b, c, d)

- q_3 is internal node-disjoint with q_1 and with q_2.
- q_1 and q_2 are not internal node-disjoint, and $\alpha_2 = \beta_2$ holds for the subpaths $\vec{\alpha_2}$ ($\vec{\beta_2}$) of $\vec{q_1}$ ($\vec{q_2}$) from the first until the last node which is contained in q_2 (q_1).
- q_1, q_2 and q_3 are internal node-disjoint with $P_u \cup P_\ell$.
- BC2a: q_1 small and $\vec{\alpha_2} = \vec{\beta_2}$; BC2b: q_1 small and $\vec{\alpha_2} \neq \vec{\beta_2}$; BC2c: q_1 large and $\vec{\alpha_2} = \vec{\beta_2}$; BC2d: q_1 large and $\vec{\alpha_2} \neq \vec{\beta_2}$.

BC3:

- q_1, q_2 and q_3 are pairwise internal node-disjoint.
- q_2 and q_3 are internal node-disjoint with $P_u \cup P_\ell$.
- q_1 is small and substitutes all commonly used edges.
- q_1 is internal node-disjoint with R_ℓ, but not with L_ℓ, and the subpath α_1 of q_1 from the start node of $\vec{q}_1$ until the last node which is contained in L_ℓ is a subpath of L_ℓ which does not contain the start node of $\vec{q}_2$.

BC4: (with subtypes a and b)

- q_3 is internal node-disjoint with q_1 and with q_2.
- q_1 and q_2 are not internal node-disjoint, and $\vec{\alpha}_3 = \vec{\beta}_2$ holds for the subpaths $\vec{\alpha}_3$ $(\vec{\beta}_2)$ of $\vec{q}_1$ $(\vec{q}_2)$ from the first until the last node which is contained in q_2 (q_1).
- q_2 and q_3 are internal node-disjoint with $P_u \cup P_\ell$.
- q_1 is small and substitutes all commonly used edges.
- q_1 is internal node-disjoint with R_ℓ, but not with L_ℓ, and the subpath α_1 of q_1 from the start node of $\vec{q}_1$ until the last node which is contained in L_ℓ is a subpath of L_ℓ which does not contain the start node of $\vec{q}_2$.
- BC4a: $\vec{\alpha}_3 = \vec{\beta}_2$; BC4b: $\vec{\alpha}_3 \neq \vec{\beta}_2$.

No Bad Configurations

We know that the PBCs which occur in the Subcases PBC1–PBC13 cannot be BCs, since $(G, (s_1, t_1), (s_2, t_2))$ does not contain a BC. In the following, we define twelve "types" of PBCs which are no BCs, so-called NBCs, and show in Lemma 3.3.15 that each PBC has to be an NBC, if $(G, (s_1, t_1), (s_2, t_2))$ does not contain a BC. In terms of notation, we again use $\vec{q}_1$, $\vec{q}_2$ and $\vec{q}_3$ for the versions of q_1, q_2 and q_3 which are directed from left to right, and α_i, β_i and γ_i for subpaths of q_1, q_2 and q_3, respectively (as described on page 83).

Definition 3.3.14 (No Bad Configurations (NBCs)). We call a PBC $(P_u, P_\ell, q_1, q_2, q_3)$ an

- *NBCi*, for an $i \in \{1, 2, \dots, 12\}$, if the properties of NBCi (listed below) are fulfilled. Note that some of the NBCis have different subtypes which are also described below.
- *No Bad Configuration (NBC)*, if $(P_u, P_\ell, q_1, q_2, q_3)$ is an NBCi for at least one $i \in \{1, \dots, 12\}$.

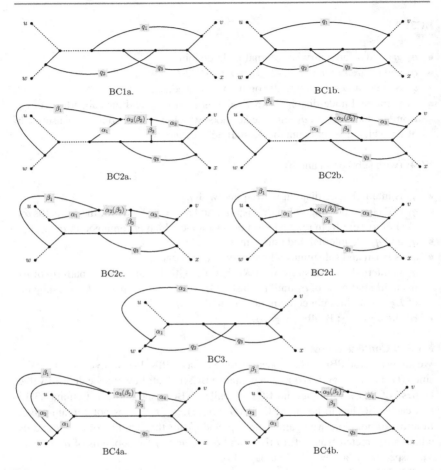

Fig. 3.41 Illustration for BCs (solid lines contain at least one edge, whereas dashed lines may consist of only one node)

For an illustration of all different NBCs, see Figure 3.42. According to the illustration, note that in some subfigures, if an alternative is not relevant for the corresponding NBCi, this alternative is not displayed (e.g., q_1 in NBC2).

NBC1:

- q_1 is small.
- q_1 and q_3 are not internal node-disjoint.

NBC2:

- q_2 and q_3 are not internal node-disjoint.

NBC3: (with subtypes a–g)

- q_2 contains a node r of P_u which is only used by the upper player.
- The different subtypes distinguish between q_1 small or large, $r \in L_u$ or $r \in R_u$, and r is substituted by q_1, or not: NBC3a: small, L_u; NBC3b: small, R_u, not substituted; NBC3c: small, R_u, substituted;
  NBC3d: large, L_u, not substituted; NBC3e: large, L_u, substituted; NBC3f: large, R_u, not substituted; NBC3g: large, R_u, substituted.

NBC4: (with subtypes a–g)

- q_3 contains a node r of P_u which is only used by the upper player.
- The different subtypes distinguish between q_1 small or large, $r \in L_u$ or $r \in R_u$, and r is substituted by q_1, or not: NBC4a: small, L_u; NBC4b: small, R_u, not substituted; NBC4c: small, R_u, substituted;
  NBC4d: large, L_u, not substituted; NBC4e: large, L_u, substituted; NBC4f: large, R_u, not substituted; NBC4g: large, R_u, substituted.

NBC5: (with subtypes a, b, c, d)

- q_1 contains nodes r of L_ℓ and r' of R_ℓ which are only used by the lower player.
- The different subtypes distinguish between q_1 small or large, and $\vec{q}_1$ first contains r, or r': NBC5a: q_1 small, first r; NBC5b: q_1 small, first r';
  NBC5c: q_1 large, first r; NBC5d: q_1 large, first r'.

NBC6: (with subtypes a, b)

- q_1 is small.
- q_1 contains a node r of R_ℓ which is only used by the lower player.
- NBC6a: r not substituted by q_3; NBC6b: r substituted by q_3.

NBC7:

- q_1 is large.
- q_1 contains a node of R_ℓ which is not substituted by q_3.

NBC8: (with subtypes a, b, c, d)

- q_1 is large.
- q_1 contains a node r of q_2 and a node r' of R_ℓ which is only used by the lower player and is substituted by q_3, *or*
  q_1 contains a node $\bar{r}$ of q_3 and a node $\bar{r}'$ of L_ℓ which is only used by the lower player and is substituted by q_2.
- The different subtypes distinguish between q_1 contains r, r' or $\bar{r}, \bar{r}'$, and $\vec{q}_1$ first contains r ($\bar{r}$) or r' ($\bar{r}'$): NBC8a: r, r' and first r; NBC8b: r, r' and first r'; NBC8c: $\bar{r}, \bar{r}'$ and first $\bar{r}$; NBC8d: $\bar{r}, \bar{r}'$ and first $\bar{r}'$.

NBC9: (with subtypes a, b)

- q_1 contains a node of L_ℓ which is not substituted by q_2.
- NBC9a: q_1 small; NBC9b: q_1 large.

NBC10:

- q_1 is small.
- q_1 contains a node r of q_2 and a node r' of L_ℓ which is only used by the lower player and is substituted by q_2, and r is before r' in $\vec{q}_1$.

NBC11: (with subtypes a, b)

- q_1 is large
- NBC11a: q_1 contains a node r of q_2 and a node r' of L_ℓ which is only used by the lower player and is substituted by q_2, and r is before r' in $\vec{q}_1$;
  NBC11b: q_1 contains a node r of q_3 and a node r' of R_ℓ which is only used by the lower player and is substituted by q_3, and r' is before r in $\vec{q}_1$.

NBC12: (with subtypes a, b)

- q_1 is large.
- q_1 contains a node r of q_2, and a node r' of q_3.
- NBC12a: r before r' in $\vec{q}_1$; NBC12b: r' before r in $\vec{q}_1$;

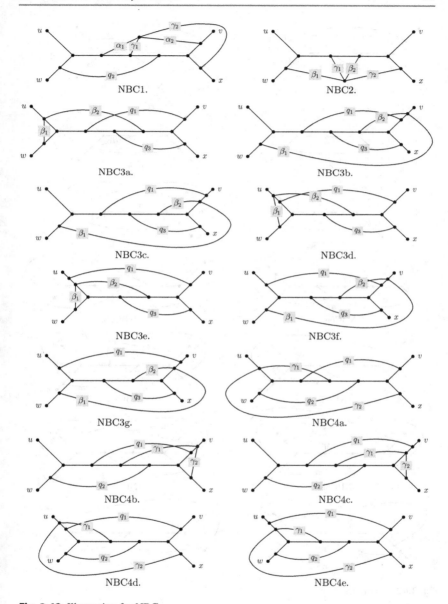

Fig. 3.42 Illustration for NBCs

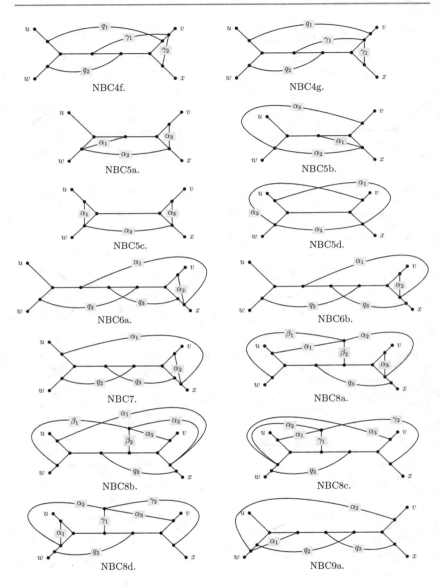

Fig. 3.42 (continued)

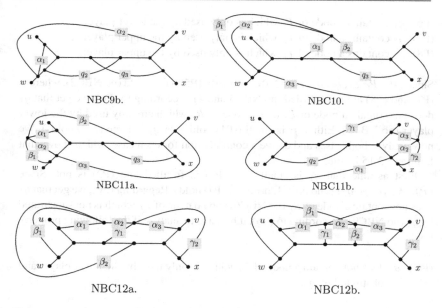

NBC9b.   NBC10.

NBC11a.   NBC11b.

NBC12a.   NBC12b.

Fig. 3.42 (continued)

The following lemma shows that the PBCs which occur in the Subcases PBC1–PBC13 have to be NBCs.

Lemma 3.3.15 *If $(G, (s_1, t_1), (s_2, t_2))$ does not contain a BC as a subgraph, then each PBC has to be an NBC.*

Proof Let us assume (by contradiction) that $(G, (s_1, t_1), (s_2, t_2))$ does not contain a BC as a subgraph, and $(P_u, P_\ell, q_1, q_2, q_3)$ is a PBC, but no NBC. The following properties are closely related to the properties defining the different types of BCs and NBCs. By investigating which of these properties are fulfilled, we construct a contradiction.

(P1) q_1 is small.
(P2) q_1 is large.
(P3) q_1 is internal node-disjoint with q_2.
(P4) q_1 is internal node-disjoint with q_3.
(P5) q_2 is internal node-disjoint with q_3.

(P6) q_1 contains a node of P_ℓ which is only used by the lower player.

(P7) q_2 contains a node of P_u which is only used by the upper player.

(P8) q_3 contains a node of P_u which is only used by the upper player.

Since $(P_u, P_\ell, q_1, q_2, q_3)$ is no NBC, property (P5) has to hold (no NBC2), whereas (P7) and (P8) are not fulfilled (no NBC3 and 4). According to (P6), we get that q_1 does not contain a node of L_ℓ *and* a node of R_ℓ which are only used by the lower player (no NBC5). Either q_1 is small ((P1) holds), or q_1 is large ((P2) holds). We now analyze both cases and derive a contradiction for each case, which shows that Lemma 3.3.15 is true.

First assume that q_1 is small, i.e., (P1) is fulfilled and (P2) is not. Since $(P_u, P_\ell, q_1, q_2, q_3)$ is no NBC1, property (P4) holds. Regarding (P6), we get that q_1 does not contain a node of R_ℓ (no NBC6) and no node of L_ℓ which is not substituted by q_2 (no NBC9). According to (P6), it remains to analyze if the following property holds:

(P6') q_1 does not contain a node of L_ℓ which is only used by the lower player and is substituted by q_2.

Furthermore we have not investigated if (P3) is fulfilled. Since $(P_u, P_\ell, q_1, q_2, q_3)$ cannot be a BC1, at least one of the properties (P3) and (P6') does not hold. We distinguish between the three cases that only (P3) does not hold, only (P6') does not hold, or both properties do not hold. In all three cases, we derive a contradiction. Assume first that (P3) does not hold, but (P6') is true. We argue that $(G, (s_1, t_1), (s_2, t_2))$ contains a BC2; contradiction. To this end, let α_2 be the subpath of q_1 from the first node r until the last node r' of $\vec{q}_1$ which is contained in q_2. Furthermore, let β_2 be the subpath of q_2 from r to r'. Now either $\alpha_2 = \beta_2$ holds, thus $(P_u, P_\ell, q_1, q_2, q_3)$ is a BC2, or $(P_u, P_\ell, q_1', q_2, q_3)$ is a BC2, where q_1' is generated by using β_2 instead of α_2 in q_1. For the case that (P6') does not hold, but (P3) is true, the argumentation is very similar. In this case, $(G, (s_1, t_1), (s_2, t_2))$ contains a BC3; contradiction: Let r be the last node of $\vec{q}_1$ which is contained in L_ℓ, only used by the lower player and substituted by q_2, and let r' be the first node of the commonly used path $P_u \cap P_\ell$ (considering it from left to right). Furthermore, let α_1 be the subpath of q_1 beginning with the start node of $\vec{q}_1$ and ending with r, and let δ be the subpath of L_ℓ from r' to r. Now either $(P_u, P_\ell, q_1, q_2, q_3)$ is a BC3, or we get a BC3 by using δ instead of α_1 in q_1. It remains to analyze the case that both properties (P3) and (P6') do not hold. Here, we will show that $(G, (s_1, t_1), (s_2, t_2))$ contains a BC4; contradiction. Let r' be the last node of $\vec{q}_1$ which is contained in L_ℓ, only used by the lower player and substituted by q_2. Furthermore, let r and $\bar{r}$ be the first, respectively last, node of $\vec{q}_1$

which is contained in q_2. Since $(P_u, P_\ell, q_1, q_2, q_3)$ is no NBC10, r' is before r in $\vec{q_1}$. Denote the subpath of q_1 from the start node of $\vec{q_1}$ to r' by α_1, and the subpaths of q_1 and q_2 from r to $\bar{r}$ by α_3 and β_2. Finally, let δ be the subpath of L_ℓ from the first node of the commonly used path $P_u \cap P_\ell$ (considering it from left to right) to r'. Now either $(P_u, P_\ell, q_1, q_2, q_3)$ is a BC4, or we get a BC4 by using δ instead of α_1, and β_2 instead of α_3, in q_1. Altogether, we derived a contradiction for the case that q_1 is small.

Now assume that q_1 is large, i.e., (P2) is fulfilled, and (P1) not. Since $(P_u, P_\ell, q_1, q_2, q_3)$ is no NBC7 and no NBC9, q_1 does not contain a node of R_ℓ which is not substituted by q_3, and q_1 does not contain a node of L_ℓ which is not substituted by q_2. It remains to investigate if (P3), (P4), (P6′) and the following property (P6″) hold:

(P6″) q_1 does not contain a node of R_ℓ which is only used by the lower player and is substituted by q_3.

Since $(P_u, P_\ell, q_1, q_2, q_3)$ is no BC1, at least one of these properties does not hold. Also note that at most one of the properties (P3) and (P4) can be violated (no NBC12), and also at most one of (P6′) and (P6″) (no NBC5). Therefore at most two properties of (P3), (P4), (P6′) and (P6″) are violated. Let us first consider the case that exactly one property is violated: If this is (P3) or (P4), we get a contradiction since there is no BC2 (either $(P_u, P_\ell, q_1, q_2, q_3)$ or $(P_u, P_\ell, q_1, q_3, q_2)$ is a BC2, or we can change q_1 to q_1' (as described in the case of q_1 small) and get that $(P_u, P_\ell, q_1', q_2, q_3)$ or $(P_u, P_\ell, q_1', q_3, q_2)$ is a BC2). If (P6′) or (P6″) does not hold, we also get a contradiction since there is no BC3 (by a suitable change of q_1 to q_1' (as described in the case of q_1 small), we get that either $(P_u, P_\ell, q_1', q_2, q_3)$ or $(P_u, P_\ell, q_1', q_3, q_2)$ is a BC3). It remains to consider the case that two properties are violated. First assume that q_1 is not internal node-disjoint with q_2 (i.e., (P4) is true, (P3) not). Since $(P_u, P_\ell, q_1, q_2, q_3)$ is no NBC8, (P6″) has to be true, and thus (P6′) is violated. Almost analogously to the case that q_1 is small, we now derive a contradiction to the fact that there is no BC4. Let r' be the last node of $\vec{q_1}$ which is contained in L_ℓ, only used by the lower player and substituted by q_2. Furthermore, let r and $\bar{r}$ be the first, respectively last, node of $\vec{q_1}$ which is contained in q_2. Since $(P_u, P_\ell, q_1, q_2, q_3)$ is no NBC11, r' is before r in $\vec{q_1}$. Denote the subpath of q_1 from the start node of $\vec{q_1}$ to r' by α_1, and the subpaths of q_1 and q_2 from r to $\bar{r}$ by α_3 and β_2. Finally, let δ be the subpath of L_ℓ from the first node of the commonly used path $P_u \cap P_\ell$ (considering it from left to right) to r'. Now we get a BC4 by using δ instead of α_1, and (possibly) β_2 instead of α_3, in q_1. We thus derived a contradiction for the case that q_1 is not internal node-disjoint with q_2. Therefore we can now assume that q_1 is internal node-disjoint with q_2, i.e., (P3) is true, which implies that (P4) is violated (two properties are violated). Since $(P_u, P_\ell, q_1, q_2, q_3)$ is no NBC8, property (P6′) is true, which implies that (P6″) is violated. By analogous argumentation as in the case that (P3) and (P6′) are violated, we now get that $(P_u, P_\ell, q_1', q_3, q_2)$ is a BC4 for a suitable change of q_1 to q_1'; contradiction. Altogether, we also derived a contradiction for the case that q_1 is large, completing the proof of Lemma 3.3.15. □

Properties of NBCs

Lemma 3.3.15 shows that the PBCs which occur in the Subcases PBC1–PBC13 have to be NBCs. Furthermore, Assumption 3.3.16 obviously is fulfilled for each occurred PBC. Therefore, we now analyze NBCs fulfilling Assumption 3.3.16 in Lemma 3.3.17–Lemma 3.3.29 (depending on the different types of NBCs). The derived properties are finally used in Lemma 3.3.30–Lemma 3.3.42 to get contradictions for the Subcases PBC1–PBC13.

Assumption 3.3.16   For a given PBC $(P_u, P_\ell, q_1, q_2, q_3)$, assume that the following properties hold:

- $P_u \cup P_\ell =: F$ is an optimal Steiner forest.
- For the strategy profile $P = (P_1, P_2)$ corresponding to F, there is an optimal solution $(\xi_{i,f})_{i \in \{1,2\}, f \in P_i}$ of LP(P) satisfying:
  - q_1 is a *tight* alternative for the upper player.
  - q_2 and q_3 are *tight* alternatives for the lower player.
  - All commonly used edges of F are completely paid.

To simplify notation in the proofs of the following Lemma 3.3.17–Lemma 3.3.29, we use path labels for the paths itself, but also for the costs of the corresponding paths (the cost of a path is the sum of edge costs over edges contained in the path). Additionally, we just write $\xi_{u,p}$ and $\xi_{\ell,p}$ for the sum of cost shares that the upper player, and the lower player, respectively, pays for the edges of a path p.

Fig. 3.43 NBC1
$(q_1 = (\alpha_1, \alpha_2),$
$q_3 = (\gamma_1, \gamma_2))$

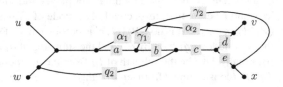

Lemma 3.3.17 *If $(P_u, P_\ell, q_1, q_2, q_3)$ is an NBC1 (cf. Figure 3.43 and Definition 3.3.14) fulfilling Assumption 3.3.16, the following properties hold:*

   i) All edges which are substituted by q_1 or by q_3 (a, b, c, d, e) are completely paid.
   ii) For $(u, v, w, x) \in \{(s_1, t_1, s_2, t_2), (t_1, s_1, t_2, s_2)\}$, Player 1 completely pays all commonly used edges which are substituted by q_1, but not by q_3 (a).
   iii) For $(u, v, w, x) \in \{(s_2, t_2, s_1, t_1), (t_2, s_2, t_1, s_1)\}$, Player 2 completely pays all commonly used edges which are substituted by q_1, but not by q_3 (a).

Proof Since q_1 is a tight alternative for the upper player, and q_3 is a tight alternative for the lower player, we get

$$\alpha_1 + \alpha_2 = \xi_{u,a} + \xi_{u,b} + \xi_{u,c} + \xi_{u,d} \text{ and } \gamma_1 + \gamma_2 = \xi_{\ell,b} + \xi_{\ell,c} + \xi_{\ell,e}.$$

Adding these equalities yields (using that all commonly used edges are completely paid)

$$\alpha_1 + \alpha_2 + \gamma_1 + \gamma_2 = \xi_{u,a} + b + c + \xi_{u,d} + \xi_{\ell,e}.$$

Since $P_u \cup P_\ell$ is an optimal Steiner forest, $\alpha_1 + \alpha_2 + \gamma_2 \geq a + b + c + d + e$ holds: Adding α_1, α_2 and γ_2 to $P_u \cup P_\ell$, while deleting a, b, c, d and e, yields an edge set containing a Steiner forest, and the cost of this Steiner forest is at least the cost of $P_u \cup P_\ell$. Altogether this yields $\xi_{u,a} + \xi_{u,d} + \xi_{\ell,e} \geq a + d + e$ and therefore $\xi_{u,a} = a, \xi_{u,d} = d$ and $\xi_{\ell,e} = e$ holds. □

Fig. 3.44 NBC2 $(q_2 = (\beta_1, \beta_2), q_3 = (\gamma_1, \gamma_2))$

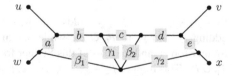

Lemma 3.3.18 *If $(P_u, P_\ell, q_1, q_2, q_3)$ is an NBC2 (cf. Figure 3.44 and Definition 3.3.14) fulfilling Assumption 3.3.16, the following properties hold:*

   i) For $(u, v, w, x) \in \{(s_1, t_1, s_2, t_2), (t_1, s_1, t_2, s_2)\}$, there is a tight alternative of Player 2 which substitutes all edges of P_2 which are substituted by q_2 or q_3 $(a, b, c, d, e;$ in particular all commonly used edges). Furthermore, Player 2 *completely pays all commonly used edges which are substituted by q_2 and q_3 (c).*

ii) *For $(u, v, w, x) \in \{(s_2, t_2, s_1, t_1), (t_2, s_2, t_1, s_1)\}$, there is a tight alternative of Player 1 which substitutes all edges of P_1 which are substituted by q_2 or q_3 (a, b, c, d, e; in particular all commonly used edges). Furthermore, Player 1 completely pays all commonly used edges which are substituted by q_2 and q_3 (c).*

Proof Since q_2 and q_3 are tight alternatives for the lower player, we get

$$\beta_1 + \beta_2 = \xi_{\ell,a} + \xi_{\ell,b} + \xi_{\ell,c} \text{ and } \gamma_1 + \gamma_2 = \xi_{\ell,c} + \xi_{\ell,d} + \xi_{\ell,e}.$$

Adding these two equalities yields

$$\beta_1 + \beta_2 + \gamma_1 + \gamma_2 = \xi_{\ell,a} + \xi_{\ell,b} + 2\xi_{\ell,c} + \xi_{\ell,d} + \xi_{\ell,e}.$$

Furthermore, the union of the paths β_1 and γ_2 contains an alternative q for the lower player which substitutes a, b, c, d and e. Using that the cost shares are a feasible $LP(P)$-solution, we can thus conclude

$$\beta_1 + \gamma_2 \geq q \geq \xi_{\ell,a} + \xi_{\ell,b} + \xi_{\ell,c} + \xi_{\ell,d} + \xi_{\ell,e}.$$

Using this, we get

$$\xi_{\ell,a} + \xi_{\ell,b} + 2\xi_{\ell,c} + \xi_{\ell,d} + \xi_{\ell,e} = \beta_1 + \beta_2 + \gamma_1 + \gamma_2 \geq \xi_{\ell,a} + \xi_{\ell,b} + \xi_{\ell,c} + \xi_{\ell,d} + \xi_{\ell,e} + \beta_2 + \gamma_1,$$

thus $\beta_2 + \gamma_1 \leq \xi_{\ell,c} \leq c$ holds. On the other hand, we get $c \leq \beta_2 + \gamma_1$, since adding β_2 and γ_1 to $P_u \cup P_\ell$, while deleting c, yields an edge set containing a Steiner forest, and the cost of this Steiner forest cannot be smaller than the cost of $P_u \cup P_\ell$. This implies $\beta_2 + \gamma_1 = \xi_{\ell,c} = c$. Furthermore, we get that q is *tight*, since $\xi_{\ell,a} + \xi_{\ell,b} + 2\xi_{\ell,c} + \xi_{\ell,d} + \xi_{\ell,e} = \beta_1 + \gamma_2 + \xi_{\ell,c}$, and therefore $\xi_{\ell,a} + \xi_{\ell,b} + \xi_{\ell,c} + \xi_{\ell,d} + \xi_{\ell,e} = \beta_1 + \gamma_2 = q$ holds. $\qquad\square$

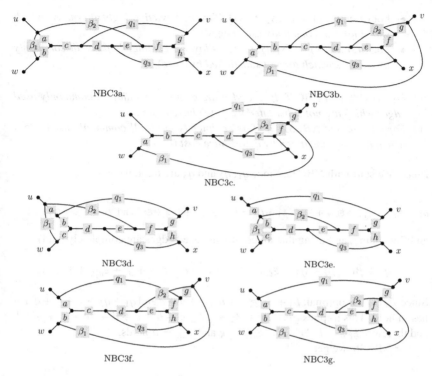

Fig. 3.45 NBC3 ($q_2 = (\beta_1, \beta_2)$)

Lemma 3.3.19   *If* $(P_u, P_\ell, q_1, q_2, q_3)$ *is an NBC3 (cf. Figure 3.45 and Definition 3.3.14) fulfilling Assumption 3.3.16, the following properties hold:*

i) *For NBC3a, NBC3d and NBC3e: All edges which are substituted by* q_1, q_2 *or* q_3 *(b, c, d, e, f, g, h for NBC3a and NBC3d; a, b, c, d, e, f, g, h for NBC3e) are completely paid.*

ii) *For NBC3a and* $(u, v, w, x) = (t_1, s_1, t_2, s_2)$*: Player 2 completely pays all commonly used edges which are not substituted by* q_1 *(c).*

iii) *For NBC3b, NBC3c, NBC3f and NBC3g:*

- *For* $(u, v, w, x) = (s_1, t_1, s_2, t_2)$*, Player 2 has a tight left alternative which substitutes all commonly used edges.*

- For $(u, v, w, x) = (t_1, s_1, t_2, s_2)$, *Player 2 has a tight right alternative which substitutes all commonly used edges.*
- For $(u, v, w, x) = (s_2, t_2, s_1, t_1)$, *Player 2 completely pays all commonly used edges which are not substituted by q_2 (e).*

iv) *For NBC3b and NBC3f: If f and g are completely paid, all commonly used edges which are not substituted by q_2 (e) have cost 0.*

v) *For NBC3c and NBC3g: If f is completely paid, all commonly used edges which are not substituted by q_2 (e) have cost 0.*

Proof We start with NBC3a. Since q_1, q_2 and q_3 are tight, we get

$$q_1 = \xi_{u,d} + \xi_{u,e} + \xi_{u,f} + \xi_{u,g}, \; \beta_1 + \beta_2 = \xi_{\ell,b} + \xi_{\ell,c} + \xi_{\ell,d} + \xi_{\ell,e} \text{ and } q_3 = \xi_{\ell,e} + \xi_{\ell,f} + \xi_{\ell,h}$$

and (by adding, and using that all commonly used edges are completely paid)

$$q_1 + \beta_1 + \beta_2 + q_3 = \xi_{\ell,b} + \xi_{\ell,c} + d + e + \xi_{\ell,e} + f + \xi_{u,g} + \xi_{\ell,h}.$$

Since $P_u \cup P_\ell$ is optimal, $\beta_1 + \beta_2 \geq a + b + c + d + e$ and $q_1 + q_3 \geq e + f + g + h$ has to hold. This implies $q_1 + \beta_1 + \beta_2 + q_3 \geq a + b + c + d + 2e + f + g + h$ and therefore $\xi_{\ell,b} = b, \xi_{\ell,c} = c, \xi_{u,g} = g$ and $\xi_{\ell,h} = h$ holds.

For NBC3b, we get

$$\beta_1 + \beta_2 = \xi_{\ell,a} + \xi_{\ell,b} + \xi_{\ell,c} + \xi_{\ell,d}$$

since q_2 is a tight alternative of the lower player. The union of β_1 with g and f contains an alternative q for the lower player which substitutes a, b, c, d and e, thus

$$\beta_1 + f + g \geq q \geq \xi_{\ell,a} + \xi_{\ell,b} + \xi_{\ell,c} + \xi_{\ell,d} + \xi_{\ell,e}$$

holds. Furthermore we get $\beta_2 \geq f + g$ since $P_u \cup P_\ell$ is optimal. Together this implies

$$\xi_{\ell,a} + \xi_{\ell,b} + \xi_{\ell,c} + \xi_{\ell,d} = \beta_1 + \beta_2 \geq \beta_1 + f + g \geq q \geq \xi_{\ell,a} + \xi_{\ell,b} + \xi_{\ell,c} + \xi_{\ell,d} + \xi_{\ell,e}$$

and therefore $\xi_{\ell,e} = 0$, q is tight and $\beta_2 = f + g$ holds. We now use $\beta_2 \geq \xi_{u,e} + \xi_{u,f} + \xi_{u,g}$ (see Lemma 3.3.20 for a proof of this inequality) to get

$$f + g = \beta_2 \geq \xi_{u,e} + \xi_{u,f} + \xi_{u,g} = e + \xi_{u,f} + \xi_{u,g}.$$

If $\xi_{u,f} = f$ and $\xi_{u,g} = g$ holds, we conclude that $e = 0$.

The analysis of NBC3c is very similar to the analysis of NBC3b: Using that q_2 is tight yields

$$\beta_1 + \beta_2 = \xi_{\ell,a} + \xi_{\ell,b} + \xi_{\ell,c} + \xi_{\ell,d}.$$

The union of β_1 with f contains an alternative q for the lower player which substitutes a, b, c, d and e, thus

$$\beta_1 + f \geq q \geq \xi_{\ell,a} + \xi_{\ell,b} + \xi_{\ell,c} + \xi_{\ell,d} + \xi_{\ell,e}$$

holds. Furthermore we get $\beta_2 \geq f$ since $P_u \cup P_\ell$ is optimal. Together this implies

$$\xi_{\ell,a} + \xi_{\ell,b} + \xi_{\ell,c} + \xi_{\ell,d} = \beta_1 + \beta_2 \geq \beta_1 + f \geq q \geq \xi_{\ell,a} + \xi_{\ell,b} + \xi_{\ell,c} + \xi_{\ell,d} + \xi_{\ell,e}$$

and therefore $\xi_{\ell,e} = 0$, q is tight and $\beta_2 = f$ holds. Lemma 3.3.20 yields $\beta_2 \geq \xi_{u,e} + \xi_{u,f}$, and we get

$$f = \beta_2 \geq \xi_{u,e} + \xi_{u,f} = e + \xi_{u,f}.$$

If $\xi_{u,f} = f$ holds, we conclude that $e = 0$.

Now we consider NBC3d. Since q_1, q_2 and q_3 are tight, and since all commonly used edges are completely paid, we get (by adding)

$$q_1 + \beta_1 + \beta_2 + q_3 = \xi_{u,b} + \xi_{\ell,c} + d + e + \xi_{\ell,e} + f + \xi_{u,g} + \xi_{\ell,h}.$$

Since $P_u \cup P_\ell$ is optimal, $\beta_1 + \beta_2 \geq a + b + c + d + e$ and $q_1 + q_3 \geq b + e + f + g + h$, and therefore $q_1 + \beta_1 + \beta_2 + q_3 \geq a + 2b + c + d + 2e + f + g + h$ holds. This implies $\xi_{u,b} = b = 0$, $\xi_{\ell,c} = c$, $\xi_{u,g} = g$ and $\xi_{\ell,h} = h$.

The properties for NBC3e follow almost analogously to the properties of NBC3d: Using that q_1, q_2 and q_3 are tight and all commonly used edges are completely paid, we get

$$q_1 + \beta_1 + \beta_2 + q_3 = \xi_{u,a} + \xi_{u,b} + \xi_{\ell,c} + d + e + \xi_{\ell,e} + f + \xi_{u,g} + \xi_{\ell,h}.$$

Since $P_u \cup P_\ell$ is optimal, $\beta_1 + \beta_2 \geq b + c + d + e$ and $q_1 + q_3 \geq a + b + e + f + g + h$, and therefore $q_1 + \beta_1 + \beta_2 + q_3 \geq a + 2b + c + d + 2e + f + g + h$ holds. This implies $\xi_{u,a} = a$, $\xi_{u,b} = b = 0$, $\xi_{\ell,c} = c$, $\xi_{u,g} = g$ and $\xi_{\ell,h} = h$.

It remains to analyze the cases NBC3f and NBC3g. The only difference between NBC3f and NBC3b, respectively NBC3g and NBC3c, is that q_1 is large in NBC3f and NBC3g, and small in NBC3b and NBC3c. In the analysis of NBC3b and NBC3c, we did not use the fact that q_1 is small. Thus, the properties for NBC3f and NBC3g follow from the analysis of NBC3b and NBC3c. $\square$

Lemma 3.3.20   *In the situation of Lemma 3.3.19, assume that* $(P_u, P_\ell, q_1, q_2, q_3)$ *is an NBC3b, NBC3c, NBC3f or NBC3g. Then* $\beta_2 \geq \xi_{u,b} + \xi_{u,c}$ *holds, where* $q_2 = (\beta_1, \beta_2)$ *and the paths b and c are as displayed in Figure 3.46.*

Fig. 3.46 Illustration for
Lemma 3.3.20

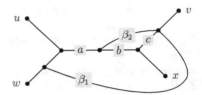

Proof Recall that β_2 is internal node-disjoint with P_ℓ, since $q_2 = (\beta_1, \beta_2)$ is an alternative for the lower player. If β_2 is an alternative for the upper player, it is clear that $\beta_2 \geq \xi_{u,b} + \xi_{u,c}$ holds. Therefore assume that β_2 is not an alternative for the upper player, i.e., it is not internal node-disjoint with P_u.

First consider the case that β_2 contains a node of L_u. Note that each node of β_2 which is contained in L_u is only used by the upper player, since β_2 is internal node-disjoint with P_ℓ. Let r be the last node in $\vec{\beta}_2$ which is contained in L_u (recall that $\vec{\beta}_2$ denotes the version of β_2 which is directed from left to right). Furthermore, let q be the subpath of β_2 from r to the first node after r of $\vec{\beta}_2$ which is contained in R_u. Then, q is an alternative for the upper player which substitutes a subpath d of L_u, all commonly used edges, and a subpath e of R_u. Either q substitutes all edges of c, or not. Subdivide c into the subpaths c_1 and c_2, where c_1 contains all edges which are substituted by q (thus, c_2 may consist of only one node). Figure 3.47 illustrates the two possible cases. We get

$$q \geq \xi_{u,d} + \xi_{u,a} + \xi_{u,b} + \xi_{u,e} \geq \xi_{u,b} + \xi_{u,c_1}.$$

Furthermore, $\beta_2 \geq q + c_2$ holds: If c_2 only consists of one node, the inequality follows since q is a subpath of $\beta_2 = (\beta_2', q, \beta_2'')$ (cf. Figure 3.47); otherwise, the

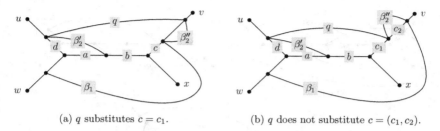

(a) q substitutes $c = c_1$.　　　　　　　(b) q does not substitute $c = (c_1, c_2)$.

Fig. 3.47 $\beta_2 = (\beta_2', q, \beta_2'')$ contains a node of L_u

subpath β_2'' fulfills $\beta_2'' \geq c_2$ due to the optimality of $P_u \cup P_\ell$, and this also implies the inequality. Using that $c_2 \geq \xi_{u,c_2}$, the desired inequality follows:

$$\beta_2 \geq q + c_2 \geq \xi_{u,b} + \xi_{u,c_1} + \xi_{u,c_2} = \xi_{u,b} + \xi_{u,c}$$

We showed the desired inequality for the case that β_2 contains a node of L_u. Thus we can now assume that β_2 does not contain a node of L_u. Since β_2 is internal node-disjoint with P_ℓ, but not internal node-disjoint with P_u, we can conclude that β_2 contains a node of R_u which is only used by the upper player, and which is different from the endnode of $\vec{\beta}_2$. Let q be the subpath of β_2 from the startnode of $\vec{\beta}_2$ until the first node of $\vec{\beta}_2$ which is contained in R_u. This is an alternative for the upper player which substitutes b and a subpath d of R_u. Again subdividing c into c_1 and c_2, where c_1 contains all edges of c which are substituted by q, yields

$$q \geq \xi_{u,b} + \xi_{u,d} \geq \xi_{u,b} + \xi_{u,c_1}.$$

As in the case above, $\beta_2 \geq q + c_2 \geq q + \xi_{u,c_2}$ holds (using that β_2 can be subdivided in β_2', q, β_2'', and that $P_u \cup P_\ell$ is optimal). This yields the desired inequality $\beta_2 \geq \xi_{u,b} + \xi_{u,c}$ and completes the proof.　　　　　　　　　　　　　　　　　　　□

Lemma 3.3.21　*If $(P_u, P_\ell, q_1, q_2, q_3)$ is an NBC4 (cf. Figure 3.48 and Definition 3.3.14) fulfilling Assumption 3.3.16, the following properties hold:*

i) For NBC4a:
- *Edges which are substituted by q_1 or q_3 (d, e, f, g, h) are completely paid.*
- *For $(u, v, w, x) = (t_1, s_1, t_2, s_2)$, Player 2 has a tight left alternative which substitutes all commonly used edges.*

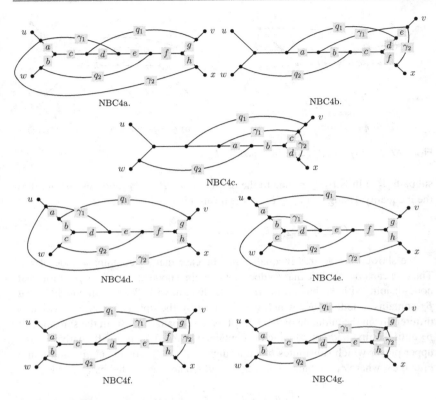

Fig. 3.48 NBC4 ($q_3 = (\gamma_1, \gamma_2)$)

- • *For $(u, v, w, x) = (t_2, s_2, t_1, s_1)$, Player 2 completely pays all commonly used edges which are not substituted by q_3 (c,d).*
- ii) *For NBC4b:*
  - • *Edges which are substituted by q_1 or q_3 (a, b, c, d, f) are completely paid.*
  - • *All commonly used edges which are substituted by q_3 (b, c) have cost 0.*
- iii) *For NBC4c:*
  - • *Edges which are substituted by q_3 (a, b, d) are completely paid.*
  - • *For $(u, v, w, x) \in \{(s_1, t_1, s_2, t_2), (t_1, s_1, t_2, s_2)\}$, Player 2 completely pays all commonly used edges which are substituted by q_3 (a, b).*

- *For $(u, v, w, x) \in \{(s_2, t_2, s_1, t_1), (t_2, s_2, t_1, s_1)\}$, Player 1 completely pays all commonly used edges which are substituted by q_3 (a, b).*

iv) *For NBC4d and NBC4e:*

- *For $(u, v, w, x) = (s_1, t_1, s_2, t_2)$, Player 2 has a tight right alternative which substitutes all commonly used edges.*
- *For $(u, v, w, x) = (s_2, t_2, s_1, t_1)$, Player 1 has a tight right alternative which substitutes all commonly used edges.*
- *For $(u, v, w, x) = (t_2, s_2, t_1, s_1)$, Player 2 completely pays all edges of the commonly used path which are not substituted by q_3 (d).*

v) *For NBC4f and NBC4g: All edges which are substituted by q_1, q_2 or q_3 (a, b, c, d, e, f, h for NBC4f; a, b, c, d, e, f, g, h for NBC4g) are completely paid.*

Proof We start with NBC4a. Since q_3 is a tight alternative of the lower player,

$$\gamma_1 + \gamma_2 = \xi_{\ell,e} + \xi_{\ell,f} + \xi_{\ell,h}$$

holds. Furthermore, the union of a and γ_2 contains an alternative q for the lower player which substitutes c, d, e, f and h. This implies

$$a + \gamma_2 \geq q \geq \xi_{\ell,c} + \xi_{\ell,d} + \xi_{\ell,e} + \xi_{\ell,f} + \xi_{\ell,h}.$$

Using $\gamma_1 \geq a$ (from the optimality of $P_u \cup P_\ell$) yields

$$\xi_{\ell,e} + \xi_{\ell,f} + \xi_{\ell,h} = \gamma_1 + \gamma_2 \geq a + \gamma_2 \geq q \geq \xi_{\ell,c} + \xi_{\ell,d} + \xi_{\ell,e} + \xi_{\ell,f} + \xi_{\ell,h} \geq \xi_{\ell,e} + \xi_{\ell,f} + \xi_{\ell,h},$$

thus q is tight and $\xi_{\ell,c} = \xi_{\ell,d} = 0$ holds. Since q_1 is tight, all commonly used edges are completely paid, and $P_u \cup P_\ell$ is optimal, we get

$$d + e + f + \xi_{u,g} + \xi_{\ell,h} = q_1 + \gamma_1 + \gamma_2 \geq q_1 + \gamma_2 \geq d + e + f + g + h.$$

This yields $\xi_{u,g} = g$ and $\xi_{\ell,h} = h$.

For NBC4b, using that q_3 is tight and $P_u \cup P_\ell$ is optimal yields

$$\xi_{\ell,b} + \xi_{\ell,c} + \xi_{\ell,f} = \gamma_1 + \gamma_2 \geq b + c + d + e + f.$$

Therefore, $\xi_{\ell,b} = b, \xi_{\ell,c} = c, \xi_{u,d} = d = 0$ and $\xi_{\ell,f} = f$ holds. Furthermore we get $b = c = 0$, since

$$\xi_{u,a} = \xi_{u,a} + \xi_{u,b} + \xi_{u,c} + \xi_{u,d} = q_1 \geq a + b + c.$$

follows from the fact that q_1 is tight.

For NBC4c, we get $\xi_{\ell,a} = a$, $\xi_{\ell,b} = b$ and $\xi_{\ell,d} = d$ by using

$$\xi_{\ell,a} + \xi_{\ell,b} + \xi_{\ell,d} = \gamma_1 + \gamma_2 \geq a + b + c + d$$

(q_3 is tight; $P_u \cup P_\ell$ is optimal).

The properties for NBC4d, NBC4e, NBC4f and NBC4g follow from the properties of the corresponding cases of NBC3 (by symmetry). □

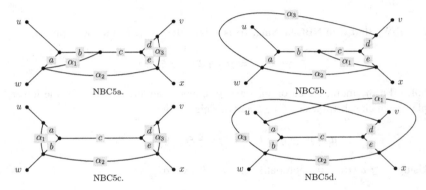

Fig. 3.49 NBC5 ($q_1 = (\alpha_1, \alpha_2, \alpha_3)$)

Lemma 3.3.22 *If $(P_u, P_\ell, q_1, q_2, q_3)$ is an NBC5 (cf. Figure 3.49 and Definition 3.3.14) fulfilling Assumption 3.3.16, the following properties hold:*

i) *All edges which are substituted by q_1 (c, d for q_1 small; a, c, d for q_1 large) are completely paid.*

ii) *For $(u, v, w, x) \in \{(s_1, t_1, s_2, t_2), (t_1, s_1, t_2, s_2)\}$, all commonly used edges are completely paid by Player 1.*

iii) *For $(u, v, w, x) \in \{(s_2, t_2, s_1, t_1), (t_2, s_2, t_1, s_1)\}$, all commonly used edges are completely paid by Player 2.*

Proof. For NBC5a and NBC5b, we get $\alpha_1 + \alpha_2 + \alpha_3 = \xi_{u,c} + \xi_{u,d}$ since q_1 is tight. The optimality of $P_u \cup P_\ell$ implies $\alpha_2 \geq b + c$ and $\alpha_3 \geq d$ and therefore $\alpha_2 + \alpha_3 \geq b + c + d$. Altogether we get $\xi_{u,c} = c, \xi_{u,d} = d$ and $\xi_{u,b} = b = 0$.

For NBC5c and NBC5d, we use $\alpha_1 + \alpha_2 + \alpha_3 = \xi_{u,a} + \xi_{u,c} + \xi_{u,d}$, since q_1 is tight, and $\alpha_1 \geq a, \alpha_2 \geq c$ and $\alpha_3 \geq d$ since $P_u \cup P_\ell$ is optimal, to get $\xi_{u,a} = a, \xi_{u,c} = c$ and $\xi_{u,d} = d$. $\qquad\square$

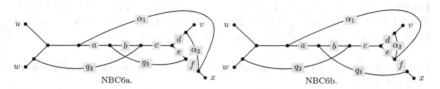

NBC6a.   NBC6b.

Fig. 3.50 NBC6 $(q_1 = (\alpha_1, \alpha_2))$

Lemma 3.3.23 *If $(P_u, P_\ell, q_1, q_2, q_3)$ is an NBC6 (cf. Figure 3.50 and Definition 3.3.14) fulfilling Assumption 3.3.16, the following properties hold:*

i) *Edges which are substituted by q_1 or by q_3 (a, b, c, d, e for NBC6a; a, b, c, d, e, f for NBC6b) are completely paid.*

ii) *For $(u, v, w, x) \in \{(s_1, t_1, s_2, t_2), (t_1, s_1, t_2, s_2)\}$, Player 1 completely pays all commonly used edges which are substituted by q_1 (a, b, c).*

iii) *For $(u, v, w, x) \in \{(s_2, t_2, s_1, t_1), (t_2, s_2, t_1, s_1)\}$, Player 2 completely pays all commonly used edges which are substituted by q_1 (a, b, c).*

Proof Since q_1 is tight and $P_u \cup P_\ell$ is optimal, we get

$$\xi_{u,a} + \xi_{u,b} + \xi_{u,c} + \xi_{u,d} = \alpha_1 + \alpha_2 \geq a + b + c + d + e.$$

Therefore $\xi_{u,a} = a, \xi_{u,b} = b, \xi_{u,c} = c, \xi_{u,d} = d$ and $\xi_{\ell,e} = e = 0$ holds. It remains to show that $\xi_{\ell,f} = f$ holds for NBC6b. This can be seen by using the above properties, the fact that q_3 is tight, and that $P_u \cup P_\ell$ is optimal:

$$\xi_{\ell,f} = \xi_{\ell,b} + \xi_{\ell,c} + \xi_{\ell,e} + \xi_{\ell,f} = q_3 \geq e + f = f.$$

$\qquad\square$

Fig. 3.51 NBC7
$(q_1 = (\alpha_1, \alpha_2))$

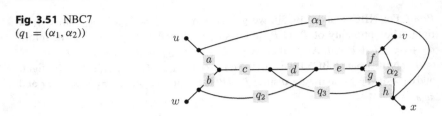

Lemma 3.3.24 *If $(P_u, P_\ell, q_1, q_2, q_3)$ is an NBC7 (cf. Figure 3.51 and Definition 3.3.14) fulfilling Assumption 3.3.16, we get that all edges which are substituted by q_1, q_2 or by q_3 (a, b, c, d, e, f, g) are completely paid.*

Proof Using that q_1, q_2 and q_3 are tight, and all commonly used edges are completely paid, yields

$$\alpha_1 + \alpha_2 + q_2 + q_3 = \xi_{u,a} + \xi_{\ell,b} + c + d + \xi_{\ell,d} + e + \xi_{u,f} + \xi_{\ell,g}.$$

The optimality of $P_u \cup P_\ell$ implies $\alpha_1 + q_2 \geq a+b+c+d$ and $\alpha_2 + q_3 \geq d+e+f+g$ and therefore

$$\alpha_1 + \alpha_2 + q_2 + q_3 \geq a + b + c + 2d + e + f + g.$$

Altogether we get $\xi_{u,a} = a, \xi_{\ell,b} = b, \xi_{u,f} = f$ and $\xi_{\ell,g} = g$. $\square$

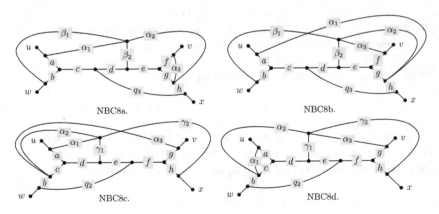

Fig. 3.52 NBC8 $(q_1 = (\alpha_1, \alpha_2, \alpha_3), q_2 = (\beta_1, \beta_2), q_3 = (\gamma_1, \gamma_2))$

Lemma 3.3.25 *If* $(P_u, P_\ell, q_1, q_2, q_3)$ *is an NBC8 (cf. Figure 3.52 and Definition 3.3.14) fulfilling Assumption 3.3.16, the following properties hold:*

i) *For NBC8a and NBC8b:*
- *All edges which are substituted by* q_1 *or by* q_2 (a, b, c, d, e, f) *are completely paid.*
- *For* $(u, v, w, x) = (s_2, t_2, s_1, t_1)$, *Player 2 completely pays all commonly used edges which are not substituted by* q_2 (e).

ii) *For NBC8c and NBC8d:*
- *All edges which are substituted by* q_1 *or by* q_3 (a, d, e, f, g, h) *are completely paid.*
- *For* $(u, v, w, x) = (t_2, s_2, t_1, s_1)$, *Player 2 completely pays all commonly used edges which are not substituted by* q_3 (d).

Proof We show the stated properties for NBC8a and NBC8b in detail. The properties for NBC8c and NBC8d follow analogously (by symmetry). For NBC8a and NBC8b, we get

$$\alpha_1 + \alpha_2 + \alpha_3 + \beta_1 + \beta_2 = \xi_{u,a} + \xi_{\ell,b} + c + d + \xi_{u,e} + \xi_{u,f}$$

since q_1 and q_2 are tight alternatives, and all commonly used edges are completely paid. On the other hand, the optimality of $L_u \cup L_\ell$ implies

$$\alpha_1 + \alpha_2 + \alpha_3 + \beta_1 \geq a + b + c + d + e + f + g$$

and therefore $\xi_{u,a} = a, \xi_{\ell,b} = b, \xi_{u,e} = e$ and $\xi_{u,f} = f$ holds. $\square$

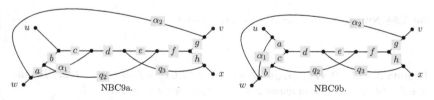

Fig. 3.53 NBC9 $(q_1 = (\alpha_1, \alpha_2))$

Lemma 3.3.26 *If $(P_u, P_\ell, q_1, q_2, q_3)$ is an NBC9 (cf. Figure 3.53 and Definition 3.3.14) fulfilling Assumption 3.3.16, the following properties hold:*

i) *Edges which are substituted by q_1, q_2 or q_3 (b, c, d, e, f, g, h for NBC9a; a, c, d, e, f, g, h for NBC9b) are completely paid.*
ii) *If q_1 is small and $(u, v, w, x) = (t_1, s_1, t_2, s_2)$, Player 2 completely pays all commonly used edges which are not substituted by q_1 (c).*

Proof We show the statement for NBC9a in detail. For NBC9b, the desired property follows from the analysis of NBC7 (by symmetry). For NBC9a we get

$$\alpha_1 + \alpha_2 + q_2 + q_3 = \xi_{\ell,b} + \xi_{\ell,c} + d + e + \xi_{\ell,e} + f + \xi_{u,g} + \xi_{\ell,h}$$

since q_1, q_2 and q_3 are tight, and all commonly used edges are completely paid. On the other hand, the optimality of $L_u \cup L_\ell$ implies $\alpha_2 + q_3 \geq e + f + g + h$, $q_2 \geq c + d + e$ and $\alpha_1 \geq a + b$ and therefore

$$\alpha_1 + \alpha_2 + q_2 + q_3 \geq a + b + c + d + 2e + f + g + h.$$

Altogether we get that $\xi_{\ell,b} = b, \xi_{\ell,c} = c, \xi_{u,g} = g$ and $\xi_{\ell,h} = h$ holds. $\square$

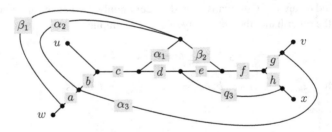

Fig. 3.54 NBC10 $(q_1 = (\alpha_1, \alpha_2, \alpha_3), q_2 = (\beta_1, \beta_2))$

Lemma 3.3.27 *If $(P_u, P_\ell, q_1, q_2, q_3)$ is an NBC10 (cf. Figure 3.54 and Definition 3.3.14) fulfilling Assumption 3.3.16, the following properties hold:*

i) *Edges which are substituted by q_1, q_2 or q_3 (a, b, c, d, e, f, g, h) are completely paid.*

ii) For $(u, v, w, x) = (t_1, s_1, t_2, s_2)$, Player 2 completely pays all commonly used edges which are not substituted by q_1 (c).

Proof Using that q_1, q_2 and q_3 are tight, and all commonly used edges are completely paid, yields

$$\alpha_1 + \alpha_2 + \alpha_3 + \beta_1 + \beta_2 + q_3 = \xi_{\ell,a} + \xi_{\ell,b} + \xi_{\ell,c} + d + e + \xi_{\ell,e} + f + \xi_{u,g} + \xi_{\ell,h}.$$

The optimality of $L_u \cup L_\ell$ yields $\alpha_1 + \beta_1 \geq a + b$, $\alpha_2 + \beta_2 \geq c + d + e$ and $\alpha_3 + q_3 \geq e + f + g + h$ and therefore

$$\alpha_1 + \alpha_2 + \alpha_3 + \beta_1 + \beta_2 + q_3 \geq a + b + c + d + 2e + f + g + h.$$

This results in $\xi_{\ell,a} = a$, $\xi_{\ell,b} = b$, $\xi_{\ell,c} = c$, $\xi_{u,g} = g$ and $\xi_{\ell,h} = h$.   □

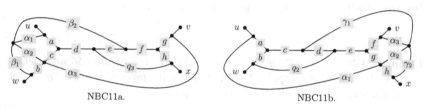

NBC11a.   NBC11b.

Fig. 3.55 NBC11 $(q_1 = (\alpha_1, \alpha_2, \alpha_3), q_2 = (\beta_1, \beta_2), q_3 = (\gamma_1, \gamma_2))$

Lemma 3.3.28 *If $(P_u, P_\ell, q_1, q_2, q_3)$ is an NBC11 (cf. Figure 3.55 and Definition 3.3.14) fulfilling Assumption 3.3.16, we get that all edges which are substituted by q_1, q_2 or by q_3 (a, b, c, d, e, f, g, h) are completely paid.*

Proof We only show the statement for NBC11a, since the properties for NBC11b follow analogously (by symmetry). For NBC11a, we get

$$\alpha_1 + \alpha_2 + \alpha_3 + \beta_1 + \beta_2 + q_3 = \xi_{u,a} + \xi_{\ell,b} + \xi_{\ell,c} + d + e + \xi_{\ell,e} + f + \xi_{u,g} + \xi_{\ell,h}$$

by using that q_1, q_2 and q_3 are tight and all commonly used edges are completely paid. On the other hand, the optimality of $L_u \cup L_\ell$ implies $\alpha_3 + q_3 \geq e + f + g + h$ and $\alpha_1 + \beta_1 + \beta_2 \geq a + b + c + d + e$ and therefore

$$\alpha_1 + \alpha_2 + \alpha_3 + \beta_1 + \beta_2 + q_3 \geq a + b + c + d + 2e + f + g + h.$$

This yields $\xi_{u,a} = a, \xi_{\ell,b} = b, \xi_{\ell,c} = c, \xi_{u,g} = g$ and $\xi_{\ell,h} = h$. □

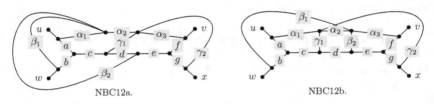

NBC12a.   NBC12b.

Fig. 3.56 NBC12 ($q_1 = (\alpha_1, \alpha_2, \alpha_3), q_2 = (\beta_1, \beta_2,), q_3 = (\gamma_1, \gamma_2)$)

Lemma 3.3.29 *If* $(P_u, P_\ell, q_1, q_2, q_3)$ *is an NBC12 (cf. Figure 3.56 and Defini-tion 3.3.14) fulfilling Assumption 3.3.16, the following properties hold if* q_1, q_2 *or* q_3 *substitutes an edge which is not completely paid:*

i) $(P_u, P_\ell, q_1, q_2, q_3)$ *is an NBC12b.*
ii) *For* $(u, v, w, x) \in \{(s_1, t_1, s_2, t_2), (t_1, s_1, t_2, s_2)\}$, *Player 2 has a positive cost share for a commonly used edge which is substituted by* q_2 *and by* q_3 *(d).*
iii) *For* $(u, v, w, x) \in \{(s_2, t_2, s_1, t_1), (t_2, s_2, t_1, s_1)\}$, *Player 1 has a positive cost share for a commonly used edge which is substituted by* q_2 *and by* q_3 *(d).*

Proof Assume that q_1, q_2 or q_3 substitutes an edge which is not completely paid. Since all commonly used edges are completely paid, this yields $\xi_{u,a} + \xi_{\ell,b} + \xi_{u,f} + \xi_{\ell,g} < a + b + f + g$.

We first show that $(P_u, P_\ell, q_1, q_2, q_3)$ is an NBC12b. Assume, on the contrary, that we have an NBC12a. Since q_1, q_2 and q_3 are tight and all commonly used edges are completely paid, we get

$$\alpha_1 + \alpha_2 + \alpha_3 + \beta_1 + \beta_2 + \gamma_1 + \gamma_2 = \xi_{u,a} + \xi_{\ell,b} + c + d + \xi_{\ell,d} + e + \xi_{u,f} + \xi_{\ell,g}.$$

Since $P_u \cup P_\ell$ is an optimal Steiner forest, we get $\alpha_1 + \beta_1 + \beta_2 \geq a + b + c + d$ and $\alpha_3 + \gamma_1 + \gamma_2 \geq d + e + f + g$ and together

$$\alpha_1 + \alpha_2 + \alpha_3 + \beta_1 + \beta_2 + \gamma_1 + \gamma_2 \geq a + b + c + 2d + e + f + g.$$

This shows that $\xi_{u,a} = a, \xi_{\ell,b} = b, \xi_{u,f} = f$ and $\xi_{\ell,g} = g$ has to hold; contradiction.

We can thus assume that $(P_u, P_\ell, q_1, q_2, q_3)$ is an NBC12b. Using that the alternatives are tight and all commonly used edges are completely paid yields

$$\alpha_1 + \alpha_2 + \alpha_3 + \beta_1 + \beta_2 + \gamma_1 + \gamma_2 = \xi_{u,a} + \xi_{\ell,b} + c + d + \xi_{\ell,d} + e + \xi_{u,f} + \xi_{\ell,g}$$

and the optimality of $P_u \cup P_\ell$ implies

$$\alpha_1 + \alpha_2 + \alpha_3 + \beta_1 + \gamma_2 \geq a + b + c + d + e + f + g.$$

Therefore $\xi_{\ell,d} \geq a + b + f + g - (\xi_{u,a} + \xi_{\ell,b} + \xi_{u,f} + \xi_{\ell,g}) > 0$ has to hold. $\square$

Analysis of Subcases PBC1–PBC13
Now we have all tools to show that the remaining subcases of Case R, i.e., Subcase PBC1–Subcase PBC13, cannot occur. As the names of these subcases already suggest, in each of these cases we derived a PBC. Since $(G, (s_1, t_1), (s_2, t_2))$ does not contain a BC as a subgraph, we can assume due to Lemma 3.3.15 that each derived PBC is an NBC. Furthermore, each PBC fulfills Assumption 3.3.16, thus we can use the properties of NBCs derived in Lemma 3.3.17–Lemma 3.3.29. For a given PBC, the proof approach is therefore to go through all different types of NBCs, and use the corresponding properties from Lemma 3.3.17–Lemma 3.3.29, as well as the properties which the PBC inherited from the corresponding subcase of Case R, to get a contradiction. As is turns out, most of the Subcases PBC1–PBC13 are quite straightforward and require very similar arguments. To prevent that the proofs get too long, we forego to repeat very similar arguments in detail (but we always refer to former lemmas containing detailed explanations).

Note that in the figures illustrating the following lemmas, we sometimes use thick lines to emphasize which alternatives are part of the considered PBC.

Lemma 3.3.30 (PBC1). *Subcase PBC1, see Figure 3.57 for illustration, cannot occur (note that it is irrelevent for the argumentation whether $\mu' < \mu''$, as illustrated, or not).*

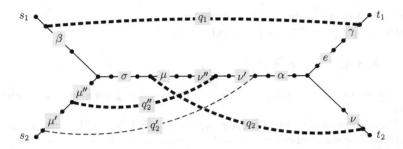

Fig. 3.57 PBC1

Proof Assume that we are in Subcase PBC1. In particular, this means the following: The alternative q_1, a smallest right tight alternative of Player 1 for e, is large. The edge e_α is the largest commonly used edge which is not completely paid by Player 2, and q_2 is a smallest right tight alternative of Player 2 for e_α. The edge e_σ is the largest edge in $\{e_{\ell_1+\ell_2+1}, \ldots, e_{\mu-1}\}$ which is not completely paid by Player 2, and q_2'' is a smallest left tight alternative of Player 2 for e_σ. Finally, $(P_1, P_2, q_1, q_2'', q_2)$ is a PBC with $(u, v, w, x) = (s_1, t_1, s_2, t_2)$.

Using Lemma 3.3.15, as well as the fact that $(G, (s_1, t_1), (s_2, t_2))$ does not contain a BC as a subgraph, we conclude that $(P_1, P_2, q_1, q_2'', q_2)$ is an NBC. We now analyze each type of NBC, and derive a contradiction if $(P_1, P_2, q_1, q_2'', q_2)$ is of this type. This shows that Subcase PBC1 cannot occur. In Table 3.2, we summarize the derived contradictions for all types of NBCs except NBC12 (this type is more complicated and is therefore analyzed separately after the table).

We now give detailed explanations how the properties in the second column of the table are derived from the definitions and properties of NBCs (given in Definition 3.3.14 and Lemma 3.3.17–Lemma 3.3.29), and why these properties contradict properties of Subcase PBC1. If $(P_1, P_2, q_1, q_2'', q_2)$ is an NBC1, 3a-c, 4a-c, 6 or 10,

Table 3.2 PBC1: Contradictions for the different types of NBCs (except NBC12)

type of NBC	contradiction
1, 3a-c, 4a-c, 6, 10	q_1 is small
2, 3f, 3g, 4d, 4e	cheaper Steiner forest (q_1 and tight alternative for Player 2)
3d, 3e, 4f, 4g, 5, 7, 8, 9, 11	e is completely paid

we get by Definition 3.3.14 that q_1 is small. This is a contradiction to the fact that q_1 is large in Subcase PBC1. Now assume that $(P_1, P_2, q_1, q_2'', q_2)$ is an NBC2, 3f-g, or 4d-e. In each of these cases, we get that Player 2 has a tight alternative which substitutes all commonly used edges (see i) of Lemma 3.3.18, iii) of Lemma 3.3.19 and iv) of Lemma 3.3.21). Together with q_1, this alternative yields a cheaper Steiner forest than F, contradicting the optimality of F. For the cases NBC3d-e, 4f-g, 5, 7, 8, 9 and 11, we get that all edges which are substituted by q_1 are completely paid (see i) of Lemma 3.3.19, v) of Lemma 3.3.21, i) of Lemma 3.3.22, Lemma 3.3.24, Lemma 3.3.25, i) of Lemma 3.3.26 and Lemma 3.3.28). Since e is substituted by q_1, this yields that e is completely paid, contradicting the fact that e is not completely paid in Subcase PBC1.

It remains to analyze the case that $(P_1, P_2, q_1, q_2'', q_2)$ is an NBC12. Since q_1 substitutes e (which is not completely paid), we get from ii) of Lemma 3.3.29 that there is an edge e_τ in $\{e_\mu, \ldots, e_{\nu''}\}$ which is not completely paid by Player 1. We now use an argumentation similar to the one used on page 75 for Subcase R.1.2.1, namely that certain changes of the cost shares do not yield a feasible solution for $LP(P)$. Here, the optimality of the given cost shares for $LP(P)$ implies that the following changes do not yield a feasible solution for $LP(P)$ for any $\varepsilon > 0$ (the sum of all collected cost shares is higher than before):

$$(**) \quad \begin{cases} \text{Decrease } \xi_{1,e_\sigma} \text{ and } \xi_{1,e_\alpha}, \text{ and increase } \xi_{1,e_\tau} \text{ and } \xi_{1,e} \text{ by } \varepsilon. \\ \text{Increase } \xi_{2,e_\sigma} \text{ and } \xi_{2,e_\alpha}, \text{ and decrease } \xi_{2,e_\tau} \text{ by } \varepsilon. \end{cases}$$

Similar as on page 75, this implies that at least one of the following alternatives (A1')–(A6') exists:

(A1') A tight alternative of Player 1 for e_τ which neither substitutes e_σ, nor e_α.
(A2') A tight alternative of Player 1 for e which does not substitute e_α.
(A3') A tight alternative of Player 1 for e which substitutes e_τ, but not e_σ.
(A4') A tight alternative of Player 2 for e_α which does not substitute e_τ.
(A5') A tight alternative of Player 2 for e_α which substitutes e_σ.
(A6') A tight alternative of Player 2 for e_σ which does not substitute e_τ.

We now argue that alternative (A5') exists, since (A1')–(A4') and (A6') are not possible: For (A1'), note that any tight alternative of Player 1 for e_τ is a left or a right alternative, and thus also substitutes e_σ or e_α (recall that if a player i does not pay a commonly used edge f completely, the existence of a tight alternative of player i for f which is neither a right nor a left alternative leads to a cheaper

Steiner forest than F; contradiction). Since q_1 is a smallest right tight alternative of Player 1 for e, the alternatives (A2$'$) and (A3$'$) cannot exist. Analogously, (A4$'$) and (A6$'$) are not possible since q_2 is a smallest right tight alternative of Player 2 for e_α, and q_2'' is a smallest left tight alternative of Player 2 for e_σ. Thus, there has to be a tight alternative of Player 2 for e_α which also substitutes e_σ. Furthermore, any such alternative is a right alternative, since a left one leads (together with q_1) to a cheaper Steiner forest than F (recall that e_α is the largest commonly used edge which is not paid completely by Player 2). Let $\bar{q}_2$, defined by $\bar{\mu}$ and $\bar{\nu}$, be a largest right tight alternative of Player 2 (in particular, $\bar{q}_2$ substitutes e_σ). If $\bar{q}_2$ substitutes all commonly used edges, or Player 2 completely pays the commonly used edges which are not substituted by $\bar{q}_2$, we get a cheaper Steiner forest than F by using q_1 and $\bar{q}_2$; contradiction. Therefore we can assume that e_ρ is the largest edge in $\{e_{\ell_1+\ell_2+1}, \ldots, e_{\bar{\mu}-1}\}$ which Player 2 does not pay completely.

Note that q_2'' is a tight left alternative of Player 2 for e_ρ. If q_2'' is a *smallest* left tight alternative of Player 2 for e_ρ, the changes of the cost shares described in (∗∗), with ρ instead of σ, lead to a feasible solution of LP(P) for some $\varepsilon > 0$; contradiction: The existence of alternatives (A1$'$)–(A4$'$) can be excluded by the same argumentation as before. If alternative (A5$'$) exists, it is a left alternative (since $\bar{q}_2$ is a *largest* right tight alternative of Player 2, there is no right tight alternative of Player 2 for e_ρ) and leads to a cheaper Steiner forest than F (together with q_1). Finally, (A6$'$) is not possible since q_2'' substitutes e_τ. We can thus assume that $\widehat{q}_2$ (defined by $\widehat{\mu}$ and $\widehat{\nu}$), a smallest left tight alternative of Player 2 for e_ρ, is *strictly smaller* than q_2''. This implies that $\widehat{q}_2$ does not substitute e_σ, since q_2'' is a smallest left tight alternative of Player 2 for e_σ. Furthermore, if $\widehat{\nu} \in \{\rho, \ldots, \bar{\mu}-1\}$ holds, we get a cheaper Steiner forest than F, and thus a contradiction, by using $q_1, \widehat{q}_2$ and $\bar{q}_2$ (note that Player 2 completely pays the edges $e_{\rho+1}, \ldots, e_{\bar{\mu}-1}$ by the choice of ρ). We can thus assume that $\widehat{\nu} \in \{\bar{\mu}, \ldots, \sigma-1\}$ holds, see Figure 3.58 for illustration (the ordering of μ'' and $\widehat{\mu}$, as well as ν and $\bar{\nu}$, is not relevant for the argumentation).

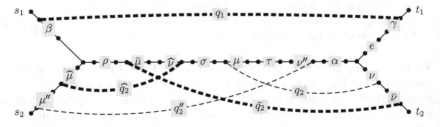

Fig. 3.58 Subcase PBC1: Remaining situation

Now, $(P_1, P_2, q_1, \widehat{q_2}, \bar{q}_2)$ is a PBC with $(u, v, w, x) = (s_1, t_1, s_2, t_2)$. By the same argumentation as before, we get that $(P_1, P_2, q_1, \widehat{q_2}, \bar{q}_2)$ is an NBC, and for all types except NBC12, we can easily derive contradictions (as given in Table 3.2). We can thus assume that $(P_1, P_2, q_1, \widehat{q_2}, \bar{q}_2)$ is an NBC12, and the edge e_ω is the largest edge in $\{e_{\bar{\mu}}, \ldots, e_{\widehat{v}}\}$ which Player 1 does not pay completely (from ii) of Lemma 3.3.29, since q_1 substitutes e which is not completely paid). We now change the cost shares for the edges e_ρ and e_ω as follows: Player 1 decreases ξ_{1,e_ρ} and increases ξ_{1,e_ω} by ε, whereas Player 2 increases ξ_{2,e_ρ} and decreases ξ_{2,e_ω} by ε. Since q_1 in particular is a smallest right tight alternative of Player 1 for e_ω, and $\widehat{q_2}$ is a smallest left tight alternative of Player 2 for e_ρ, there is an $\varepsilon > 0$ such that the described changes yield a feasible LP(P)-solution. Furthermore, by choosing $\varepsilon > 0$ small enough, we can ensure that no alternatives "get tight" (i.e., an alternative which is not tight with respect to the original cost shares, is also not tight with respect to the changed cost shares). On the other hand, q_1, q_2 and all left tight alternatives of Player 2 "stay tight" (that is, are also tight with respect to the changed cost shares). This way, we achieved that $\bar{q}_2$ is not tight anymore, whereas q_1, q_2, q_2'' and $\widehat{q_2}$ keep their properties (i.e, q_1 is a smallest right tight alternative of Player 1 for e, the alternative q_2 is a smallest right tight alternative of Player 2 for e_α, the alternative q_2'' is a smallest left tight alternative of Player 2 for e_σ, and $\widehat{q_2}$ is a smallest left tight alternative of Player 2 for e_ρ). Regarding the changes described in $(**)$, we can thus again conclude that there is a right tight alternative of Player 2 for e_α which substitutes e_σ. Furthermore, a largest such alternative is *strictly smaller* than $\bar{q}_2$, since it does not substitute e_ω. Let $\bar{\bar{q}}_2$, defined by $\bar{\bar{\mu}}$ and $\bar{\bar{v}}$, be a largest right tight alternative of Player 2 for e_σ. Regarding $\bar{\bar{\mu}}$, we get $\bar{\bar{\mu}} \in \{\omega + 1, \ldots, \sigma\}$. If $\bar{\bar{\mu}} \in \{\omega + 1, \ldots, \widehat{v}\}$ holds, $(P_1, P_2, q_1, \widehat{q_2}, \bar{\bar{q}}_2)$ is an NBC with $(u, v, w, x) = (s_1, t_1, s_2, t_2)$, and this leads to a contradiction (for all types except NBC12 see Table 3.2; NBC12 is not possible since Player 1 completely pays the edges $e_{\omega+1}, \ldots, e_{\widehat{v}}$). We can thus assume that $\bar{\bar{\mu}} \in \{\widehat{v} + 1, \ldots, \sigma\}$ holds. If Player 2 completely pays the edges $e_{\widehat{v}+1}, \ldots, e_{\bar{\bar{\mu}}-1}$, we get a contradiction since using q_1, $\widehat{q_2}$ and $\bar{\bar{q}}_2$ yields a cheaper Steiner forest than F. Thus let $e_{\rho'}$ be the largest edge in $\{e_{\widehat{v}+1}, \ldots, e_{\bar{\bar{\mu}}-1}\}$ which Player 2 does not pay completely. We can now repeat the entire argumentation of the last paragraph (starting on page 114) with $\bar{\bar{q}}_2$ and ρ' instead of $\bar{q}_2$ and ρ.

In every "iteration" of the described "procedure", the largest right tight alternative of Player 2 for e_σ gets strictly smaller. We conclude that finally, there is no right tight alternative of Player 2 for e_σ anymore, so that the changes described in $(**)$ yield a feasible LP(P)-solution for some $\varepsilon > 0$; contradiction. This completes the analysis of Subcase PBC1 and shows that this case cannot occur. $\qquad\square$

Lemma 3.3.31 (PBC2). *Subcase PBC2, see Figure 3.59 for illustration, cannot occur. According to the illustration, note that q_2' does not substitute all commonly used edges, but it may substitute e_α, but this is irrelevant for the argumentation.*

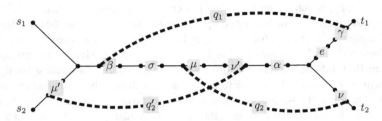

Fig. 3.59 PBC2

Proof. Assume that we are in Subcase PBC2. In particular, this means that q_1 is small, q_2' is a largest left tight alternative of Player 2 for e_σ, Player 2 does not completely pay the edge e_α, and $(P_1, P_2, q_1, q_2', q_2)$ is a PBC with $(u, v, w, x) = (s_1, t_1, s_2, t_2)$.

Lemma 3.3.15 yields that $(P_1, P_2, q_1, q_2', q_2)$ is an NBC. Analyzing the different types yields the contradictions given in Table 3.3, showing that PBC2 cannot occur. We now give detailed explanations how the properties in the second column of the table are derived from the definitions and properties of NBCs (given in Definition 3.3.14 and Lemma 3.3.17–Lemma 3.3.29), and why these properties contradict properties of Subcase PBC2.

If $(P_1, P_2, q_1, q_2'', q_2)$ is an NBC3d-g, 4d-g, 7, 8, 11 or 12, we get by Definition 3.3.14 that q_1 is large. This is a contradiction to the fact that q_1 is small in Subcase PBC2. Now assume that $(P_1, P_2, q_1, q_2'', q_2)$ is an NBC1, 3a, 4a-b, 5, 6, 9 or 10. In each of these cases, we get that all edges which are substituted by q_1 are completely paid (see i) of Lemma 3.3.17, i) of Lemma 3.3.19, i) and ii) of Lemma 3.3.21, i) of Lemma 3.3.22, i) of Lemma 3.3.23, i) of Lemma 3.3.26 and i) of Lemma 3.3.27). Since e is substituted by q_1, this yields that e is completely paid, contradicting the fact that e is not completely paid in Subcase PBC1. If $(P_1, P_2, q_1, q_2'', q_2)$ is an NBC2 or 3b-c, there is a tight left alternative of Player 2 which substitutes all commonly used edges (see i) of Lemma 3.3.18 and iii) of Lemma 3.3.19). In particular, this alternative substitutes e_σ and is larger than q_2', contradicting the fact that q_2' is a largest left tight alternative of Player 2 for e_σ. Finally, for the case that $(P_1, P_2, q_1, q_2'', q_2)$ is an NBC4c, we get from iii) of Lemma 3.3.21 that Player 2 completely pays the edge e_α, but this does not hold in Subcase PBC2. $\square$

Table 3.3 PBC2: Contradictions for the different types of NBCs

type of NBC	contradiction
3d-g, 4d-g, 7, 8, 11, 12	q_1 is large
1, 3a, 4a-b, 5, 6, 9, 10	e is completely paid
2, 3b-c	larger left tight alternative than q_2'
4c	Player 2 completely pays e_α

Lemma 3.3.32   (PBC3). *Subcase PBC3, see Figure 3.60 for illustration, cannot occur. According to the illustration, note that q_2' substitutes all commonly used edges, but it may be small, and it is irrelevant for the argumentation whether $\nu < \nu'$, as displayed, or not.*

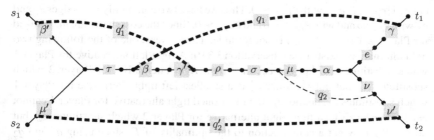

Fig. 3.60 PBC3

Proof Assume that we are in Subcase PBC3. In particular, this means that Player 2 does not completely pay the edge e_α, whereas Player 1 does not completely pay the edge e_τ (the largest edge in $\{e_{\ell_1+\ell_2+1}, \ldots, e_{\beta-1}\}$ which Player 1 does not pay completely). Furthermore, q_1' is a smallest left tight alternative for Player 1 which substitutes e_τ, but not e_ρ, and $(P_2, P_1, q_2', q_1, q_1')$ is a PBC with $(u, v, w, x) = (t_2, s_2, t_1, s_1)$.

Lemma 3.3.15 yields that $(P_2, P_1, q_2', q_1, q_1')$ is an NBC. Table 3.4 summarizes the derived contradictions for all types of NBCs (note that q_2' can be small or large) except NBC12 (this case is analyzed separately after the table). Except for the contradiction derived in NBC3b-c, 3f-g and 4b, the previous Lemma 3.3.30 and Lemma 3.3.31 contain detailed explanations for similar contradictions. Thus we only explain the case that $(P_2, P_1, q_2', q_1, q_1')$ is an NBC3b-c, 3f-g or 4b. Since e is the smallest edge which is not completely paid, we get that the edges $e_1, \ldots, e_{\ell_1}$ are completely paid. By iv) and v) of Lemma 3.3.19, and ii) of Lemma 3.3.21, we conclude that the cost of e_τ is 0. In particular, Player 1 completely pays e_τ, which is not true in Subcase PBC3.

Table 3.4 PBC3: Contradictions for the different types of NBCs (except NBC12)

type of NBC	contradiction
1, 4a, 4d-e, 5, 6, 8c-d	Player 2 completely pays e_α
2	cheaper Steiner forest (q_2' and tight alternative for Player 1)
3a, 3d-e, 4f-g, 7, 8a-b, 9, 10, 11	e is completely paid
3b-c, 3f-g, 4b	cost of e_τ is 0 (thus Player 1 completely pays e_τ)
4c	Player 1 completely pays e_τ

It remains to consider the case that $(P_2, P_1, q_2', q_1, q_1')$ is an NBC12. Since q_1 substitutes e (which is not completely paid), we get from iii) of Lemma 3.3.29 that Player 1 has a positive cost share for an edge in $\{e_\beta, \ldots, e_{\gamma'}\}$. Let e_ω be the smallest edge in $\{e_\beta, \ldots, e_{\gamma'}\}$ with $\xi_{1,e_\omega} > 0$. Thus we have two commonly used edges e_τ and e_ω with $\tau < \omega$ and with $\xi_{1,e_\omega} > 0$ and $\xi_{2,e_\tau} > 0$. Since the cost shares are maximized for Player 2, CHANGE(τ, ω) is not feasible. Thus at least one of the following two tight alternatives exists (see Observation 3.3.9): A tight left alternative for Player 1 which substitutes e_τ, but not e_ω, or a tight right alternative for Player 2 which substitutes e_ω, but not e_τ. Since q_1' is a smallest left tight alternative for Player 1 which substitutes e_τ, but not e_ρ, the mentioned tight alternative for Player 1 cannot exist. Thus let $\bar{q}_2$ be a tight right alternative for Player 2 which substitutes e_ω, but not e_τ. We now get a contradiction to the optimality of F, since using q_1 and $\bar{q}_2$ yields a cheaper Steiner forest than F (Player 1 completely pays $e_{\tau+1}, \ldots, e_{\beta-1}$ by the choice of τ, and Player 2 completely pays $e_\beta, \ldots, e_{\omega-1}$ by the choice of ω). This shows that $(P_2, P_1, q_2', q_1, q_1')$ cannot be an NBC12, and completes the proof of Lemma 3.3.32. $\qquad\square$

Lemma 3.3.33 (PBC4). *Subcase PBC4, see Figure 3.61 for illustration, cannot occur.*

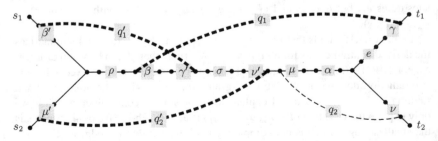

Fig. 3.61 PBC4

Proof Assume that we are in Subcase PBC4. In particular, this means that the edges e_α and e_σ are not completely paid by Player 2, whereas the edge e_ρ is not completely paid by Player 1. Moreover, the edge e_σ is the largest edge in $\{e_\beta, \ldots, e_{\mu-1}\}$ which Player 2 does not pay completely. Finally, q_2' (a largest left tight alternative of Player 2 for e_σ) is small, and $(P_2, P_1, q_2', q_1, q_1')$ is a PBC with $(u, v, w, x) = (t_2, s_2, t_1, s_1)$.

Lemma 3.3.15 yields that $(P_2, P_1, q_2', q_1, q_1')$ is an NBC. The following Table 3.5 summarizes, for each type of NBC, the derived contradiction if we assume that $(P_2, P_1, q_2', q_1, q_1')$ is of this type (for detailed explanations of very similar contradictions, see Lemma 3.3.30, Lemma 3.3.31 and Lemma 3.3.32). This shows that Subcase PBC4 cannot occur.   □

Table 3.5 PBC4: Contradictions for the different types of NBCs

type of NBC	contradiction
3d-g, 4d-g, 7, 8, 11, 12	q_2' is large
4a, 5	Player 2 completely pays e_α
2	cheaper Steiner forest (q_2, q_2' and tight alternative for Player 1)
3a, 9, 10	e is completely paid
1, 6	Player 2 completely pays e_σ
4c	Player 1 completely pays e_ρ
3b-c, 4b	cost of e_ρ is 0 (thus Player 1 completely pays e_ρ)

Lemma 3.3.34 (PBC5). *Subcase PBC5, see Figure 3.62 for illustration, cannot occur. According to the illustration, note that it is irrelevant for the argumentation whether $v < \bar{v}$, as illustrated, or not.*

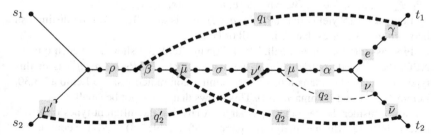

Fig. 3.62 PBC5

Proof Assume that we are in Subcase PBC5. In particular, this means that q_1, a smallest right tight alternative of Player 1 for e, is small. Furthermore, the edges e_α and e_σ are not completely paid by Player 2, and q_2' is a largest left tight alternative of Player 2 for e_σ. Finally, $(P_1, P_2, q_1, q_2', \bar{q}_2)$ is a PBC with $(u, v, w, x) = (s_1, t_1, s_2, t_2)$.

Lemma 3.3.15 yields that $(P_1, P_2, q_1, q_2', \bar{q}_2)$ is an NBC. The following Table 3.6 shows, for each type of NBC, the derived contradiction if we assume that $(P_1, P_2, q_1, q_2', \bar{q}_2)$ is of this type (for detailed explanations of very similar contradictions, see Lemma 3.3.30, Lemma 3.3.31 and Lemma 3.3.32). This shows that Subcase PBC5 cannot occur. □

Table 3.6 PBC5: Contradictions for the different types of NBCs

type of NBC	contradiction
3d-g, 4d-g, 7, 8, 11, 12	q_1 is large
1, 3a, 4a-b, 5, 6, 9, 10	e is completely paid
2	Player 2 completely pays e_σ
3b-c	larger left tight alternative than q_2'
4c	Player 2 completely pays e_α

Lemma 3.3.35 (PBC6). *Subcase PBC6, see Figure 3.63 for illustration, cannot occur. According to the illustration, note that v may also be larger than v', but this is not relevant for the argumentation.*

Proof Assume that we are in Subcase PBC6. In particular, this means that Player 2 does not completely pay the edges e_α, e_σ and e_ρ. Furthermore, there is no tight *left* alternative of Player 2 for e_σ, and q_2' is a largest right tight alternative of Player 2, whereas $\bar{q}_2$ is a smallest left tight alternative of Player 2 for e_ρ. Finally, $(P_1, P_2, q_1, \bar{q}_2, q_2')$ is a PBC with $(u, v, w, x) = (s_1, t_1, s_2, t_2)$.

Lemma 3.3.15 yields that $(P_1, P_2, q_1, \bar{q}_2, q_2')$ is an NBC. We now distinguish between the two cases that q_1 is small or large.

First assume that q_1 is small. The following Table 3.7 shows, for each type of NBC, the derived contradiction if we assume that $(P_1, P_2, q_1, \bar{q}_2, q_2')$ is of this type (for detailed explanations of very similar contradictions, see Lemma 3.3.30, Lemma 3.3.31 and Lemma 3.3.32). This shows that q_1 cannot be small.

Now consider the case that q_1 is large. Analyzing the different types of NBCs (except NBC12, which is analyzed separately after Table 3.8) yields the contradictions displayed in Table 3.8 (for detailed explanations of very similar contradictions, see Lemma 3.3.30, Lemma 3.3.31 and Lemma 3.3.32):

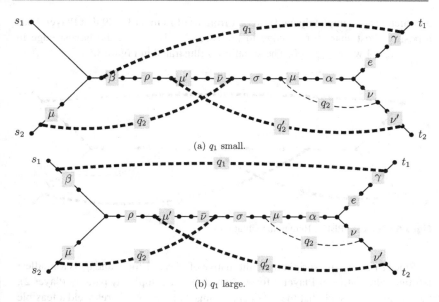

(a) q_1 small.

(b) q_1 large.

Fig. 3.63 PBC6

Table 3.7 PBC6 for q_1 small: Contradictions for the different types of NBCs

type of NBC	contradiction
3d-g, 4d-g, 7, 8, 11, 12	q_1 is large
1, 3a, 4a-b, 5, 6, 9, 10	e is completely paid
2	larger right tight alternative than q_2'
3b-c	left tight alternative of Player 2 for e_σ
4c	Player 2 completely pays e_α

Table 3.8 PBC6 for q_1 large: Contradictions for the different types of NBCs (except NBC12)

type of NBC	contradiction
1, 3a-c, 4a-c, 6, 10	q_1 is small
2, 3f-g, 4d-e	cheaper Steiner forest (q_1 and tight alternative for Player 2)
3d-e, 4f-g, 5, 7, 8, 9, 11	e is completely paid

It remains to analyze the case that $(P_1, P_2, q_1, \bar{q}_2, q_2')$ is an NBC12. We use a similar argumentation as in Lemma 3.3.30 (for the case NBC12). Since q_1 substitutes

e (which is not completely paid), we get from ii) of Lemma 3.3.29 that Player 2 has a positive cost share for an edge in $\{e_{\mu'}, \ldots, e_{\bar{v}}\}$. Let e_τ be the largest edge in $\{e_{\mu'}, \ldots, e_{\bar{v}}\}$ with $\xi_{2,e_\tau} > 0$. The situation is illustrated in Figure 3.64.

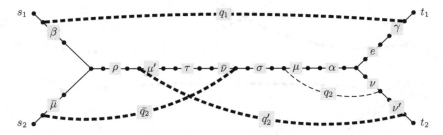

Fig. 3.64 Subcase PBC6: Remaining situation

Since q_1 is a smallest right tight alternative of Player 1 for e, and $\bar{q}_2$ is a smallest left tight alternative of Player 2 for e_ρ (which is not completely paid by Player 2), there is an $\varepsilon > 0$ such that the following changes of the cost shares yield a feasible (and optimal) solution for LP(P):

> Decrease ξ_{1,e_ρ} and increase ξ_{1,e_τ} by ε.
> Increase ξ_{2,e_ρ} and decrease ξ_{2,e_τ} by ε.

In particular, we choose $\varepsilon > 0$ small enough such that no alternatives "get tight" (i.e., an alternative which is not tight with respect to the original cost shares, also is not tight with respect to the changed cost shares). Moreover, q_1 and all tight left alternatives of Player 2 stay tight with respect to the changed cost shares. The reason why we changed the cost shares is that q'_2 is not tight anymore. If Player 2 does not have a tight alternative for e_σ anymore, there is an $\varepsilon' > 0$ such that the following changes of the cost shares yield a feasible solution for LP(P), contradicting the optimality of the current cost shares (the sum of cost shares is larger than before):

$$(*\,*\,*) \quad \begin{cases} \text{Decrease } \xi_{1,e_\sigma} \text{ and increase } \xi_{1,e} \text{ by } \varepsilon'. \\ \text{Increase } \xi_{2,e_\sigma} \text{ by } \varepsilon'. \end{cases}$$

Thus we can assume that q''_2, defined by μ'' and v'', is a largest right tight alternative of Player 2 for e_σ. Furthermore, q''_2 is strictly smaller than q'_2 since it does not substitute e_τ. If $\mu'' \in \{\tau+1, \ldots, \bar{v}\}$ holds, $(P_1, P_2, q_1, \bar{q}_2, q''_2)$ is an NBC with $(u, v, w, x) =$

(s_1, t_1, s_2, t_2) (by Lemma 3.3.15), and this leads to a contradiction (for all types except NBC12 see Table 3.8; NBC12 is not possible since Player 1 completely pays the edges $e_{\tau+1}, \ldots, e_{\bar{v}}$). We can thus assume that $\mu'' \in \{\bar{v}+1, \ldots, \sigma\}$ holds. If Player 2 completely pays the edges $e_{\bar{v}+1}, \ldots, e_{\mu''-1}$, we get a contradiction since using q_1, $\bar{q}_2$ and q_2'' yields a cheaper Steiner forest than F. Thus let $e_{\rho'}$ be the largest edge in $e_{\bar{v}+1}, \ldots, e_{\mu''-1}$ which Player 2 does not pay completely. Due to property (2M) of the cost shares, Player 2 has a tight alternative for $e_{\rho'}$. Furthermore, each such alternative is a left one, since q_2'' is a largest right tight alternative of Player 2. Let $\bar{q}_2'$, defined by $\bar{\mu}'$ and $\bar{v}'$, be a smallest left tight alternative of Player 2 for $e_{\rho'}$. If $\bar{v}' \in \{\rho', \ldots, \mu''-1\}$ holds, we get a cheaper Steiner forest than F, and thus a contradiction, by using q_1, $\bar{q}_2'$ and q_2''. Thus, $\bar{v}' \in \{\mu'', \ldots, \sigma-1\}$ holds (recall that there is no tight left alternative of Player 2 for e_σ). This implies that $(P_1, P_2, q_1, \bar{q}_2', q_2'')$ is an NBC with $(u, v, w, x) = (s_1, t_1, s_2, t_2)$. We can now repeat the entire argumentation of the case that q_1 is large (beginning on page 121) with $\bar{q}_2'$, q_2' and ρ' instead of $\bar{q}_2$, q_2' and ρ. In every "iteration" of the described "procedure", the largest right tight alternative of Player 2 for e_σ gets strictly smaller. We conclude that finally, there is no right tight alternative of Player 2 for e_σ anymore, so that the changes described in $(***)$ yield a feasible LP(P)-solution for some $\varepsilon' > 0$; contradiction. This completes the analysis of Subcase PBC6 and shows that this case cannot occur. $\square$

Lemma 3.3.36 (PBC7). *Subcase PBC7, see Figure 3.65 for illustration, cannot occur. According to the illustration, note that it is irrelevant for the argumentation whether $v < v'$, as illustrated, or not.*

Fig. 3.65 PBC7

Proof Assume that we are in Subcase PBC7. In particular, this means that $\bar{q}_2$ (a smallest left tight alternative of Player 2 for e_ρ) is small. Furthermore, the edges e_α and e_ρ are not completely paid by Player 2, whereas the edge e_ω is not completely paid by Player 1. In particular, the edge e_ρ is the largest edge in $\{e_\beta, \ldots, e_{\mu'-1}\}$

which is not completely paid by Player 2. Finally, $(P_2, P_1, \bar{q}_2, q_1, q_1')$ is a PBC with $(u, v, w, x) = (t_2, s_2, t_1, s_1)$.

Lemma 3.3.15 yields that $(P_2, P_1, \bar{q}_2, q_1, q_1')$ is an NBC. The following Table 3.9 shows, for each type of NBC, the derived contradiction if we assume that $(P_2, P_1, \bar{q}_2, q_1, q_1')$ is of this type (for detailed explanations of very similar contradictions, see Lemma 3.3.30, Lemma 3.3.31 and Lemma 3.3.32). This shows that Subcase PBC7 cannot occur.   □

Table 3.9 PBC7: Contradictions for the different types of NBCs

type of NBC	contradiction
3d-g, 4d-g, 7, 8, 11, 12	$\bar{q}_2$ is large
1, 6	Player 2 completely pays e_ρ
2	cheaper Steiner forest ($\bar{q}_2$, q_2' and tight alternative of Player 1)
3a, 9, 10	e is completely paid
3b-c, 4b	cost of e_ω is 0 (thus Player 1 completely pays e_ω)
4a, 5	Player 2 completely pays e_α
4c	Player 1 completely pays e_ω

Lemma 3.3.37 (PBC8). *Subcase PBC8, see Figure 3.66 for illustration, cannot occur. According to the illustration, note that it is irrelevant for the argumentation whether $v < v'$, as illustrated, or not.*

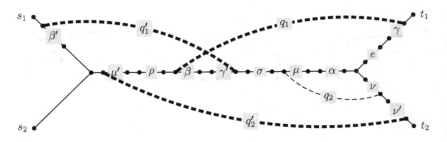

Fig. 3.66 PBC8

Proof Assume that we are in Subcase PBC8. In particular, this means that q_2' (a largest right tight alternative of Player 2 for e_σ) is small. Furthermore, the edge e_α is not completely paid by Player 2, whereas the edge e_ρ (the largest edge in

$\{e_{\mu'}, \ldots, e_{\beta-1}\}$ which Player 1 does not pay completely) is not completely paid by Player 1. The alternative q_1' is a smallest left tight alternative of Player 1 which substitutes e_ρ, but not e_σ. Furthermore, there is an edge in $\{e_\beta, \ldots, e_{\gamma'}\}$ with cost larger than zero. Finally, $(P_2, P_1, q_2', q_1', q_1)$ is a PBC with $(u, v, w, x) = (s_2, t_2, s_1, t_1)$.

Lemma 3.3.15 yields that $(P_2, P_1, q_2', q_1', q_1)$ is an NBC. Analyzing the different types of NBCs (except NBC2, which is analyzed separately after Table 3.10) yields the contradictions displayed in Table 3.10 (for detailed explanations of very similar contradictions, see Lemma 3.3.30, Lemma 3.3.31 and Lemma 3.3.32):

Table 3.10 PBC8: Contradictions for the different types of NBCs (except NBC2)

type of NBC	contradiction
3d-g, 4d-g, 7, 8, 11, 12	q_2' is large
1, 3a, 4a-c, 6, 9, 10	e is completely paid
3b, 3c, 5	Player 2 completely pays e_α

It remains to analyze the case that $(P_2, P_1, q_2', q_1', q_1)$ is an NBC2. From Lemma 3.3.18 ii) we get that Player 1 completely pays the edges $e_\beta, \ldots, e_{\gamma'}$. Recall that there is an edge in $\{e_\beta, \ldots, e_{\gamma'}\}$ with cost larger than zero. Let e_τ be the smallest edge in $\{e_\beta, \ldots, e_{\gamma'}\}$ with positive cost. This implies $\xi_{1,e_\tau} = c_{e_\tau} > 0$. Furthermore, $\xi_{2,e_\rho} > 0$ holds. Since the cost shares are maximized for Player 2, CHANGE(ρ, τ) is not feasible. Thus at least one of the following two tight alternatives exists (see Observation 3.3.9): A tight left alternative for Player 1 which substitutes e_ρ, but not e_τ, or a tight right alternative for Player 2 which substitutes e_τ, but not e_ρ. Since q_1' is a smallest left tight alternative for Player 1 which substitutes e_ρ, but not e_σ, the mentioned alternative for Player 1 cannot exist. Thus we can assume that $\bar{q}_2$ is a tight right alternative for Player 2 which substitutes e_τ, but not e_ρ. Now we get a contradiction to the optimality of F, since using $\bar{q}_2$ and q_1 yields a cheaper Steiner forest than F (Player 1 completely pays the edges $e_{\rho+1}, \ldots, e_{\beta-1}$ by the choice of ρ, and the edges $e_\beta, \ldots, e_{\tau-1}$ all have cost 0). Altogether, we conclude that PBC8 cannot occur. $\square$

Lemma 3.3.38 (PBC9). *Subcase PBC9, see Figure 3.67 for illustration, cannot occur. According to the illustration, note that it is irrelevant for the argumentation whether $v < v'$, as illustrated, or not.*

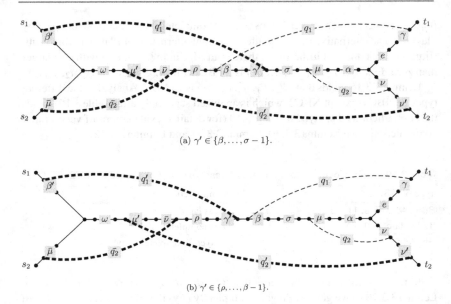

(a) $\gamma' \in \{\beta, \dots, \sigma - 1\}$.

(b) $\gamma' \in \{\rho, \dots, \beta - 1\}$.

Fig. 3.67 PBC9

Proof Assume that we are in Subcase PBC9. In particular, this means that q_1' (a smallest left tight alternative of Player 1 which substitutes e_ρ, but not e_σ) is small. Furthermore, the edges e_α and e_ω are not completely paid by Player 2, whereas the edge e_ρ is not completely paid by Player 1. Moreover, in case that $\gamma' \in \{\beta, \dots, \sigma - 1\}$, the edges $e_\beta, \dots, e_{\gamma'}$ all have cost 0, and if $\gamma' \in \{\rho, \dots, \beta - 1\}$, Player 1 completely pays the edges $e_{\gamma'+1}, \dots, e_{\beta-1}$ (since e_ρ is the largest edge in $\{e_{\mu'}, \dots, e_{\beta-1}\}$ which Player 1 does not pay completely). Finally, $(P_1, P_2, q_1', q_2', \bar{q}_2)$ is a PBC with $(u, v, w, x) = (t_1, s_1, t_2, s_2)$.

Lemma 3.3.15 yields that $(P_1, P_2, q_1', q_2', \bar{q}_2)$ is an NBC. The following Table 3.11 shows, for each type of NBC, the derived contradiction if we assume that $(P_1, P_2, q_1', q_2', \bar{q}_2)$ is of this type (for detailed explanations of very similar contradictions, see Lemma 3.3.30, Lemma 3.3.31 and Lemma 3.3.32). This shows that Subcase PBC9 cannot occur. $\square$

Table 3.11 PBC9: Contradictions for the different types of NBCs

type of NBC	contradiction
3d-g, 4d-g, 7, 8, 11, 12	q_1' is large
1, 5, 6	Player 1 completely pays e_ρ
2, 3b-c, 4a	cheaper Steiner forest (q_1', q_1 and tight alternative of Player 2)
3a, 9, 10	Player 2 completely pays e_α
4b	cost of e_ω is 0 (thus Player 2 completely pays e_ω)
4c	Player 2 completely pays e_ω

Lemma 3.3.39 (PBC10). *Subcase PBC10, see Figure 3.68 for illustration, cannot occur. According to the illustration, note that the ordering of $\widetilde{\mu}$ and $\bar{\mu}$, as well as ν, $\widehat{\nu}$ and ν', is not relevant for the argumentation. Furthermore, the exact location of $\bar{\nu} \in \{\rho, \ldots, \sigma - 1\}$ is not relevant.*

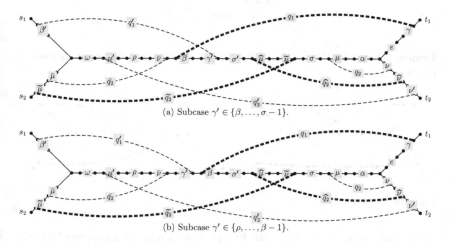

(a) Subcase $\gamma' \in \{\beta, \ldots, \sigma - 1\}$.

(b) Subcase $\gamma' \in \{\rho, \ldots, \beta - 1\}$.

Fig. 3.68 PBC10

Proof Assume that we are in Subcase PBC10. In particular, this means that q_1, a smallest right tight alternative of Player 1 for e, is small. Furthermore, the edge e_α is not completely paid by Player 2, and q_2' is a largest right tight alternative of Player 2. Moreover, in case that $\gamma' \in \{\beta, \ldots, \sigma - 1\}$, the edges $e_\beta, \ldots, e_{\gamma'}$ all have cost 0, and if $\gamma' \in \{\rho, \ldots, \beta - 1\}$, Player 1 completely pays the edges $e_{\gamma'+1}, \ldots, e_{\beta-1}$ (since

e_ρ is the largest edge in $\{e_{\mu'}, \ldots, e_{\beta-1}\}$ which Player 1 does not pay completely).
Finally, $(P_1, P_2, q_1, \widetilde{q_2}, \widehat{q_2})$ is a PBC with $(u, v, w, x) = (s_1, t_1, s_2, t_2)$.

Lemma 3.3.15 yields that $(P_1, P_2, q_1, \widetilde{q_2}, \widehat{q_2})$ is an NBC. The following Table 3.12 shows, for each type of NBC, the derived contradiction if we assume that $(P_1, P_2, q_1, \widetilde{q_2}, \widehat{q_2})$ is of this type (for detailed explanations of very similar contradictions, see Lemma 3.3.30, Lemma 3.3.31 and Lemma 3.3.32). This shows that Subcase PBC10 cannot occur.   □

Table 3.12 PBC10: Contradictions for the different types of NBCs

type of NBC	contradiction
3d-g, 4d-g, 7, 8, 11, 12	q_1 is large
1, 3a, 4a-b, 5, 6, 9, 10	e is completely paid
2	larger right tight alternative than q_2'
3b-c	cheaper Steiner forest (q_1', q_1 and tight alternative of Player 2)
4c	Player 2 completely pays e_α

Lemma 3.3.40 (PBC11). *Subcase PBC11, see Figure 3.69 for illustration, cannot occur.*

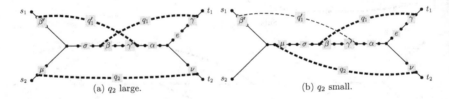

(a) q_2 large.   (b) q_2 small.

Fig. 3.69 PBC11

Proof Assume that we are in Subcase PBC11. In particular, this means that Player 2 does not completely pay the edge e_α (the largest commonly used edge which Player 2 does not pay completely). The alternative q_1' is a smallest left tight alternative of Player 1 which substitutes e_σ, but not e_α, where e_σ is the largest commonly used edge which is substituted by q_2, but not by q_1, and which Player 1 does not pay completely. Finally, $(P_2, P_1, q_2, q_1', q_1)$ is a PBC with $(u, v, w, x) = (s_2, t_2, s_1, t_1)$.

Lemma 3.3.15 yields that $(P_2, P_1, q_2, q_1', q_1)$ is an NBC. We now distinguish between the two cases that q_2 (a smallest right tight alternative of Player 2 for e_α) is large or small.

First consider the case that q_2 is large. For all types except NBC12 (this case is analyzed separately after Table 3.13), the following Table 3.13 shows the derived contradiction if we assume that $(P_2, P_1, q_2, q_1', q_1)$ is of this type (for detailed explanations of very similar contradictions, see Lemma 3.3.30, Lemma 3.3.31 and Lemma 3.3.32).

Table 3.13 PBC11 for q_2 large: Contradictions for the different types of NBCs (except NBC12)

type of NBC	contradiction
1, 3a-c, 4a-c, 6, 10	q_2 is small
2, 4d-e	cheaper Steiner forest (q_2 and tight alternative of Player 1)
3d-e, 4f-g, 7, 8c-d, 9, 11	e is completely paid
3f-g, 5, 8a-b	Player 2 completely pays e_α

It remains to consider the case that $(P_2, P_1, q_2, q_1', q_1)$ is an NBC12. Since q_1 substitutes e, which is not completely paid, we get from iii) of Lemma 3.3.29 that there is an edge $e_\tau \in \{e_\beta, \ldots, e_{\gamma'}\}$ with $\xi_{1,e_\tau} > 0$. Furthermore, $\xi_{2,e_\sigma} > 0$ holds. Since the cost shares are maximized for Player 2, CHANGE(σ, τ) is not feasible. Thus at least one of the following two tight alternatives exists (see Observation 3.3.9): A tight left alternative for Player 1 which substitutes e_σ, but not e_τ, or a tight right alternative for Player 2 which substitutes e_τ, but not e_σ. But due to our choices of q_1' and q_2, both such alternatives cannot exist; contradiction. This completes the analysis of the case that q_2 is large.

Now consider the case that q_2 is small. For all types except NBC2 (this case is analyzed separately after Table 3.14), the following Table 3.14 shows the derived contradiction if we assume that $(P_2, P_1, q_2, q_1', q_1)$ is of this type (for detailed explanations of very similar contradictions, see Lemma 3.3.30, Lemma 3.3.31 and Lemma 3.3.32).

Table 3.14 PBC11 for q_2 small: Contradictions for the different types of NBCs (except NBC2)

type of NBC	contradiction
3d-g, 4d-g, 7, 8, 11, 12	q_2 is large
1, 3a, 4a-c, 6, 9, 10	e is completely paid
3b-c, 5	Player 2 completely pays e_α

To complete the proof, it remains to consider the case that $(P_2, P_1, q_2, q_1', q_1)$ is an NBC2 (and q_2 is small). From ii) of Lemma 3.3.18, we get that Player 1 has a tight right alternative $\bar{q}_1$ for e which is large. Let q_2' (defined by μ' and ν') be a largest right tight alternative of Player 2. If q_2' substitutes all commonly used edges, or Player 2 completely pays all commonly used edges which are not substituted by q_2', we get a cheaper Steiner forest than F, and thus a contradiction, by using $\bar{q}_1$ and q_2'. Thus we can assume that e_τ is the largest edge in $\{e_{\ell_1+\ell_2+1}, \ldots, e_{\mu'-1}\}$ which Player 2 does not pay completely. Since the cost shares are maximized for Player 2, there is a tight alternative of Player 2 for e_τ. Furthermore, any such alternative is a left one, since a right one would be larger than q_2'. Let $\bar{q}_2$, defined by $\bar{\mu}$ and $\bar{\nu}$, be a smallest left tight alternative of Player 2 for e_τ. If $\bar{\nu} \leq \mu' - 1$, or $\bar{\nu} \geq \alpha$, we get a contradiction to the optimality of F: Using $\bar{q}_1$, $\bar{q}_2$ (and q_2' in the first case) yields a cheaper Steiner forest than F, since Player 2 completely pays the edges $e_{\tau+1}, \ldots, e_{\mu'-1}$ and $e_{\alpha+1}, \ldots, e_{\ell_1+\ell_2+m}$ by the choices of τ and α. Therefore, $\bar{\nu} \in \{\mu', \ldots, \alpha - 1\}$ holds. We distinguish between $\bar{\nu} \in \{\mu', \ldots, \sigma - 1\}$ and $\bar{\nu} \in \{\sigma, \ldots, \alpha - 1\}$, and derive contradictions for both cases, completing the proof of Lemma 3.3.40.

First consider the case that $\bar{\nu} \in \{\mu', \ldots, \sigma - 1\}$. In Figure 3.70, we illustrated the situation, where the exact location of $\bar{\nu} \in \{\mu', \ldots, \sigma - 1\}$ (in particular whether $\bar{\nu} < \mu$ (as illustrated) or $\bar{\nu} \geq \mu$), as well as the ordering of ν and ν', is irrelevant for the argumentation.

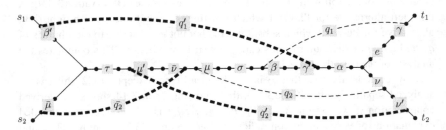

Fig. 3.70 Subcase PBC11: Situation for $\bar{\nu} \in \{\mu', \ldots, \sigma - 1\}$

We get that $(P_1, P_2, q_1', q_2', \bar{q}_2)$ is a PBC with $(u, v, w, x) = (t_1, s_1, t_2, s_2)$. Lemma 3.3.15 yields that $(P_1, P_2, q_1', q_2', \bar{q}_2)$ is an NBC. For each type, we get a contradiction (given in Table 3.15) if we assume that $(P_1, P_2, q_1', q_2', \bar{q}_2)$ is of this type, which completes the analysis of the case that $\bar{\nu} \in \{\mu', \ldots, \sigma - 1\}$.

It remains to derive a contradiction for the case that $\bar{\nu} \in \{\sigma, \ldots, \alpha - 1\}$. We use the same argumentation as in Subcase R.1.2.1 on page 75, namely that the following changes of the cost shares do not yield a feasible solution for LP(P), for any $\varepsilon > 0$ (since the sum of cost shares of Player 2 over the commonly used edges is larger

Table 3.15 PBC11 for q_2 small and $\bar{\nu} \in \{\mu', \ldots, \sigma - 1\}$: Contradictions for the different types of NBCs

type of NBC	contradiction
3d-g, 4d-g, 7, 8, 11, 12	q_1' is large
1, 5, 6	Player 1 completely pays e_σ
2, 3b, 3c, 4a	cheaper Steiner forest ($\bar{q}_1$ and tight alternative of Player 2)
3a, 9, 10	Player 2 completely pays e_α
4b	cost of e_τ is 0 (thus Player 2 completely pays e_τ)
4c	Player 2 completely pays e_τ

than before, see property (2M)):

$$\text{Increase } \xi_{1,e_\sigma} \text{ and decrease } \xi_{1,e_\tau} \text{ and } \xi_{1,e_\alpha} \text{ by } \varepsilon.$$
$$\text{Decrease } \xi_{2,e_\sigma} \text{ and increase } \xi_{2,e_\tau} \text{ and } \xi_{2,e_\alpha} \text{ by } \varepsilon.$$

This implies that at least one of the following four alternatives exists:

A tight alternative of Player 1 for e_σ which neither substitutes e_τ, nor e_α.
A tight alternative of Player 2 for e_τ which does not substitute e_σ.
A tight alternative of Player 2 for e_α which does not substitute e_σ.
A tight alternative of Player 2 which substitutes e_α and e_τ.

For each of the four alternatives, we now derive a contradiction if this alternative exists. This completes the proof of Lemma 3.3.40. First of all, we can assume that any of the four alternatives is a right or a left alternative, or both (recall that if a player i does not pay a commonly used edge f completely, the existence of a tight alternative of player i for f which is neither a right nor a left alternative leads to a cheaper Steiner forest than F). This immediately shows that the first alternative cannot exist, since it is neither a right, nor a left alternative. If the second alternative exists, this alternative is smaller than $\bar{q}_2$, but $\bar{q}_2$ is a smallest left tight alternative of Player 2 for e_τ. The third alternative contradicts the fact that q_2 is a smallest right tight alternative of Player 2 for e_α. Finally, the fourth alternative needs to be a left one, since q_2' is a largest right tight alternative of Player 2, and thus leads (together with $\bar{q}_1$) to a cheaper Steiner forest than F (since Player 2 completely pays the edges $e_{\alpha+1}, \ldots, e_{\ell_1+\ell_2+m}$). $\qquad\square$

Lemma 3.3.41 (PBC12). *Subcase PBC12, see Figure 3.71 for illustration, cannot occur. According to the illustration, note that $\bar{\nu} \in \{\mu, \ldots, \sigma - 1\}$ is also possible,* *but this is irrelevant for the argumentation, as well as it is irrelevant if $v < v'$, or not.*

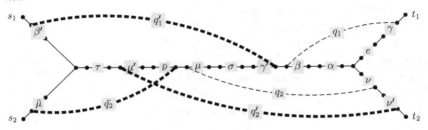

Fig. 3.71 PBC12

Proof Assume that we are in Subcase PBC12. In particular, this means that q_1', a smallest left tight alternative of Player 1 which substitutes e_σ, but not e_α, is small. Furthermore, the edges e_α and e_τ are not completely paid by Player 2, whereas e_σ is the largest edge in $\{e_\mu, \ldots, e_{\beta-1}\}$ which Player 1 does not pay completely. Finally, $(P_1, P_2, q_1', q_2', \bar{q}_2)$ is a PBC with $(u, v, w, x) = (t_1, s_1, t_2, s_2)$.

Lemma 3.3.15 yields that $(P_1, P_2, q_1', q_2', \bar{q}_2)$ is an NBC. The following Table 3.16 shows, for each type of NBC, the derived contradiction if we assume that $(P_1, P_2, q_1', q_2', \bar{q}_2)$ is of this type (for detailed explanations of very similar contradictions, see Lemma 3.3.30, Lemma 3.3.31 and Lemma 3.3.32). This shows that Subcase PBC12 cannot occur.   □

Table 3.16 PBC12: Contradictions for the different types of NBCs

type of NBC	contradiction
3d-g, 4d-g, 7, 8, 11, 12	q_1' is large
1, 5, 6	Player 1 completely pays e_σ
2, 3b-c, 4a	cheaper Steiner forest (q_1', q_1 and tight alternative of Player 2)
3a, 9, 10	Player 2 completely pays e_α
4b	cost of e_τ is 0 (thus Player 2 completely pays e_τ)
4c	Player 2 completely pays e_τ

Lemma 3.3.42 (PBC13). *Subcase PBC13, see Figure 3.72 for illustration, cannot occur.*

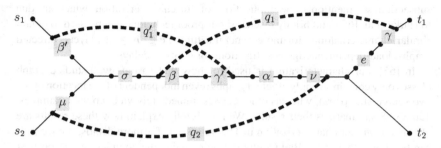

Fig. 3.72 PBC13

Proof Assume that we are in Subcase PBC13. In particular, this means that q_2 (a left tight alternative of Player 2 for e_α) is small. Furthermore, the edge e_α is the largest commonly used edge which is not completely paid by Player 2, and Player 1 does not completely pay the edge e_σ. Finally, $(P_2, P_1, q_2, q_1, q_1')$ is a PBC with $(u, v, w, x) = (t_2, s_2, t_1, s_1)$.

Lemma 3.3.15 yields that $(P_2, P_1, q_2, q_1, q_1')$ is an NBC. The following Table 3.17 shows, for each type of NBC, the derived contradiction if we assume that $(P_2, P_1, q_2, q_1, q_1')$ is of this type (for detailed explanations of very similar contradictions, see Lemma 3.3.30, Lemma 3.3.31 and Lemma 3.3.32). This shows that Subcase PBC13 cannot occur. □

Table 3.17 PBC13: Contradictions for the different types of NBCs

type of NBC	contradiction
3d-g, 4d-g, 7, 8, 11, 12	q_2 is large
1, 5, 6, 4a	Player 2 completely pays e_α
2	cheaper Steiner forest (q_2 and tight alternative of Player 1)
3a, 9, 10	e is completely paid
3b-c, 4b	cost of e_σ is 0 (thus Player 1 completely pays e_σ)
4c	Player 1 completely pays e_σ

3.3.6   Discussion of the Result

We presented a characterization of (strongly) efficient graphs for *two* player games in *undirected* graphs with *constant* edge cost functions and *zero* delays. In this subsection, we mention some implications of our characterization, which are due to Harks et al. [65], and briefly discuss if our proof technique can be used to obtain similar characterizations for more general settings, e.g., $n \geq 2$ players, directed graphs, load-dependent edge cost functions, or nonzero delays.

In [65], our characterization is used to show that various well-studied graph classes only contain strongly efficient graphs (even independent of the location of the two source-sink pairs), whereas other classes immediately yield counterexamples. Table 3.18 summarizes their results. We now briefly explain how these results are obtained from our characterization in Theorem 3.3.1 (for more details, see [65]). It can be shown that every Bad Configuration contains the complete graph on four vertices K_4 as a minor (in [65], this is shown explicitly for BC1a). Therefore, a K_4-minor-free graph cannot contain a BC as a subgraph, and is therefore strongly efficient. Note that important examples for K_4-minor-free graphs are generalized series-parallel graphs, trees, graphs containing exactly one cycle, and cactus graphs (connected graphs, such that any two cycles have at most one common vertex). Thus, these classes only contain strongly efficient graphs. A wheel (fan) is a cycle (path) together with an additional vertex which is connected to all vertices of the cycle (path). Since fans are subgraphs of wheels, it is sufficient to show that wheels do not contain Bad Configurations as subgraphs. The proof approach is to derive a contradiction, if a wheel contains a Bad Configuration. In [65], this is done explicitly for BC1a. Finally, since all Bad Configurations contain a cycle with length at least 7, and all Bad Configurations contain at least 9 edges and at least 7 vertices, graphs where the longest cycle has length at most 6, and graphs with at most 8 edges or at most 6 vertices, do not contain a Bad Configuration as a subgraph, and are thus strongly efficient. On the other hand, Figure 3.73 shows a BC1a which is planar and bipartite (the partition of the vertices is indicated by large and small vertices). Furthermore, chordal graphs contain non-efficient graphs, as complete graphs are chordal and obviously contain Bad Configurations as subgraphs.

Table 3.18 Efficiency of graph classes

(strongly) efficient classes	classes containing non-efficient graphs
K_4-minor-free graphs	planar graphs
wheel and fan graphs	bipartite graphs
longest cycle ≤ 6	chordal graphs
≤ 8 edges, or ≤ 6 vertices	

Fig. 3.73 A BC1a which is planar and bipartite

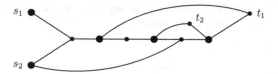

As a further consequence of our characterization, [65] obtain a lower bound of $\frac{15}{14}$ for the *worst-case* PoS resulting from an optimal solution of the system designer's problem in network cost sharing games with two players, constant shareable edge cost functions, and zero delays. To see this, consider the instance displayed in Figure 3.74, where $\delta > 0$. Note that the graph is a BC1a. It is shown in [65] that the unique optimal strategy profile (consisting of all solid edges, thus with cost $14\delta + 8$) is not enforceable, and all other profiles have cost at least $15\delta + 8$. Therefore, the ratio of the costs of a best enforceable, and an optimal profile, is at least $\frac{15\delta+8}{14\delta+8}$. As δ tends to infinity, we achieve a lower bound for the worst-case PoS of $\frac{15}{14}$. Since any worst-case instance for the PoS needs to contain a Bad Configuration as a subgraph, it is conjectured in [65] that the worst-case PoS is exactly $\frac{15}{14}$.

Fig. 3.74 A lower bound for the worst-case PoS

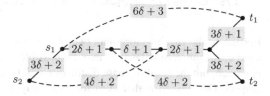

Now turn to the question whether our proof approach can be used to obtain a similar characterization for more general settings, that is, more than two players, directed graphs, more general edge costs, or nonzero delays. A key argument of our approach is that there is a "prototype" of an optimal, but not enforceable strategy profile: We can assume without loss of generality that the used edges F induce a cycle-free subgraph, and there are commonly used edges (see Subsection 3.3.5). Thus, F looks essentially like displayed in Figure 3.75a, and can be partitioned into a left, middle and right part. For directed graphs, edge cost functions which depend on the number of users, or nonzero delays, it is no longer valid in general that there is an optimal strategy profile inducing no cycles. Figure 3.75b shows an instance with a directed graph, where the only possible strategy profile induces several (directed) cycles. Thus, for directed graphs, load-dependent edge cost functions, or nonzero

delays, there is no simple prototype of an optimal strategy profile, and it is not clear
how to apply our approach.

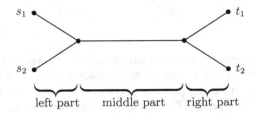

left part   middle part   right part

(a) Key argument of our approach: Protoype and partition of F.

(b) The unique strategy profile induces several cycles.

Fig. 3.75 Discussion of our approach for more general settings

Regarding the case of more than two players (in undirected graphs with constant
edge costs and zero delays), it may be possible to derive a result using our proof
approach, but the resulting case distinction will be significantly larger than for two
players. One reason is that even for three players, we need to consider different
prototypes of an optimal strategy profile (see Figure 3.76), and furthermore, the
prototypes need to be partitioned into more than three parts.

In the next section, we present a result for $n \geq 2$ players which holds for nondecreasing, discrete-concave edge cost functions and nonzero delays in undirected or
directed graphs having a special *series-parallel structure* (see Section 3.4).

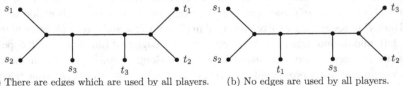

(a) There are edges which are used by all players.   (b) No edges are used by all players.

Fig. 3.76 Different prototypes of an optimal strategy profile for three players

3.4   Results for n-Player Games

In this section, we consider network cost sharing games with $n \geq 2$ players in so-called n-series-parallel graphs (which we introduced in [60]). Our main interest is to find sufficient conditions on the network to guarantee the existence of an enforceable optimal strategy profile.

In Corollary 3.4.19, we show that if $(G, (s_i, t_i)_{i \in N})$ is n-series-parallel (as defined in Definition 3.4.1), then, for all vectors c of nondecreasing and discrete-concave edge cost functions and all vectors d of player-specific delays, *every* optimal strategy profile of the network $(G, (s_i, t_i)_{i \in N}, c, d)$ is enforceable. Or, in other words, an n-series-parallel graph $(G, (s_i, t_i)_{i \in N})$ is strongly efficient for nondecreasing and discrete-concave edge costs and arbitrary delays. Recall that an edge cost function c_e is *discrete-concave* if it satisfies $c_e(x + \delta) - c_e(x) \geq c_e(y + \delta) - c_e(y)$ for all $x, y, \delta \in \mathbb{N}$ with $x \leq y$. Furthermore note that discrete-concave edge costs contain constant edge costs as a special case. Our proof of Corollary 3.4.19 is constructive: In Subsection 3.4.1, we present Algorithm n-SEPA which takes as input an arbitrary strategy profile P. If P is enforceable, the algorithm outputs $P' := P$. If P is not enforceable, the algorithm transforms P into an enforceable strategy profile P' with smaller cost, i.e., $C(P') < C(P)$. Obviously, this yields that every optimal strategy profile is enforceable: Using an optimal, but not enforceable strategy profile P as input for Algorithm n-SEPA leads to the existence of a strategy profile P' with strictly smaller cost than P, contradicting P's optimality.

We now describe further results which we obtained from Algorithm n-SEPA. As certificate for the enforceability of the output profile P', the algorithm also computes an optimal solution for LP(P') with property (BB). Using Theorem 3.2.1, we get that P' is enforceable, and we also obtain a budget-balanced and separable cost sharing method ξ such that P' is a PNE of the game induced by ξ. This gives rise to the following procedure to compute a cost sharing method with PoS equal to 1: Compute an optimal strategy profile P, and then apply Algorithm n-SEPA with P as input. But in general, computing an optimal strategy profile is $\mathcal{NP}$-hard, since the *uncapacitated facility location problem*, which is known as $\mathcal{NP}$-hard to solve, is a special case (see Subsection 3.4.4). Thus, the described procedure does not lead to the *efficient* computation of a cost sharing method with PoS of 1 in general.[10] To achieve that the described procedure is efficient, it is furthermore necessary that Algorithm n-SEPA has polynomial running time. In Theorem 3.4.20, we show that this is fulfilled for the case of constant edge cost functions. Unfortunately, the com-

---

[10]Here, we call an algorithm efficient or polynomial time if its running time is bounded by a polynomial in the encoding length of the network $\mathcal{N} = (G, (s_i, t_i)_{i \in N}, c, d)$.

putation of an optimal strategy profile remains $\mathcal{NP}$-hard for this special case (see Subsection 3.4.4). But, by combining Algorithm n-SEPA with any polynomial-time α-approximation for the underlying optimization problem of computing an optimal strategy profile, we can compute efficiently a cost sharing method with PoS at most α (for the case of constant edge costs). Thus, Algorithm n-SEPA can be seen as a black-box which reduces the efficient computation of a cost sharing method with low PoS to the design of an approximation algorithm with low approximation guarantee for the underlying optimization problem. In particular, for the special case of constant edge costs and *zero* delays in *undirected* n-series-parallel graphs, we can efficiently compute a cost sharing method with PoS equal to 1, since computing an optimal strategy profile is possible in polynomial time for this case (see Proposition 3.4.22).

We now define the class of n-series-parallel graphs that we consider in this section[11].

Definition 3.4.1 (*n*-series-parallel graph). Let $G = (V, E)$ be a directed or undirected graph with source-sink pairs $(s_i, t_i)_{i \in N}$. For $i \in N$, let G_i be the subgraph of G which is induced by the set of (s_i, t_i)-paths $\mathcal{P}_i$. We say that $(G, (s_i, t_i)_{i \in N})$ is *n-series-parallel* if for all $i \in N$, the graph G_i is *series-parallel*, that is, it can be created by a sequence of *series* and/or *parallel* operations starting from

- the undirected edge $s_i - t_i$, if G is undirected,
- the directed edge $s_i \to t_i$, if G is directed.

Here, *series* and *parallel* operations are defined as follows: For an undirected edge $e = u - v$, a series operations replaces this edge by a new vertex w and two undirected edges $u - w, w - v$; a parallel operation adds to e a parallel edge $e' = u - v$. If $e = u \to v$ is a directed edge, a series operations replaces this edge by a new vertex w and two directed edges $u \to w, w \to v$; a parallel operation adds to e a parallel edge $e' = u \to v$.

For illustration, Figure 3.77 shows an example of an undirected 3-series-parallel graph $(G, (s_i, t_i)_{i \in N = \{1,2,3\}})$, and Figure 3.78 shows constructions for the corresponding subgraphs G_i for $i \in \{1, 2, 3\}$.

---

[11]Note that in [60], we used a stronger definition of undirected n-series parallel graphs, since we additionally required that $(G, (s_i, t_i)_{i \in N})$ is *irredundant*, meaning that, for every node/edge of G, there is a player $i \in N$ and an (s_i, t_i)-path P_i containing that node/edge. But in fact, all results continue to hold also without this restriction, thus we do not require it here.

Fig. 3.77 An undirected
3-series-parallel graph
$(G, (s_i, t_i)_{i \in \{1,2,3\}})$

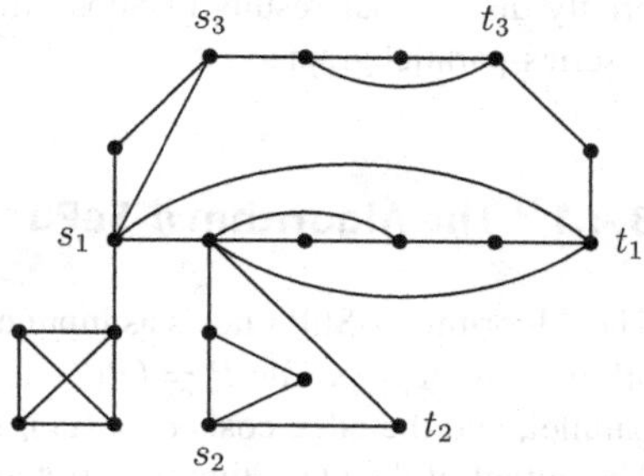

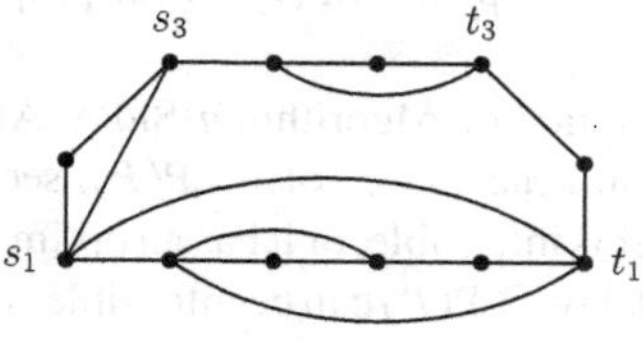

The subgraph $G_1 = G_3$. 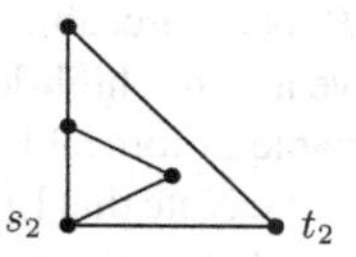   The subgraph G_2.

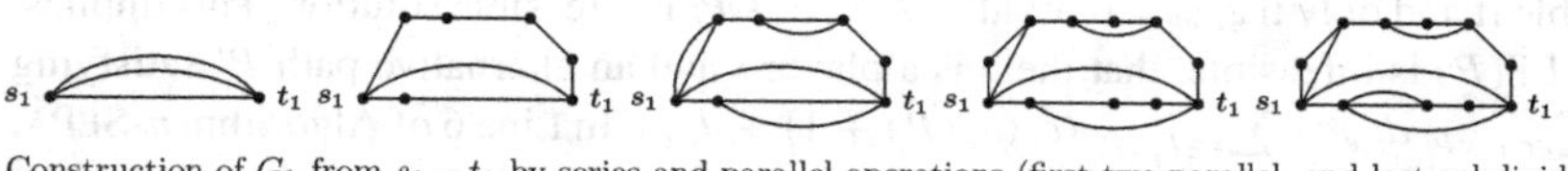

Construction of G_1 from $s_1 - t_1$ by series and parallel operations (first two parallel, and last subdivide operation(s) are left out).

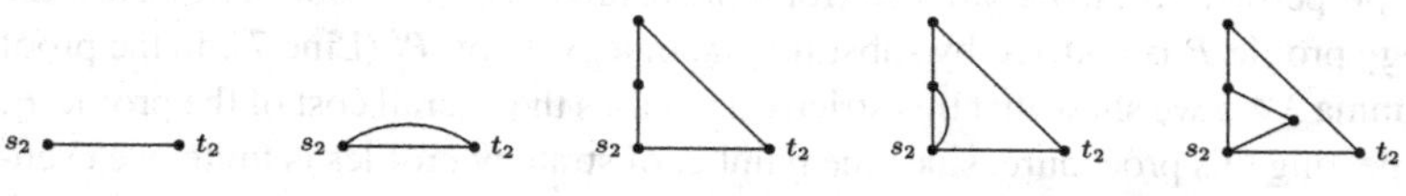

Construction of G_2 from $s_2 - t_2$ by series and parallel operations.

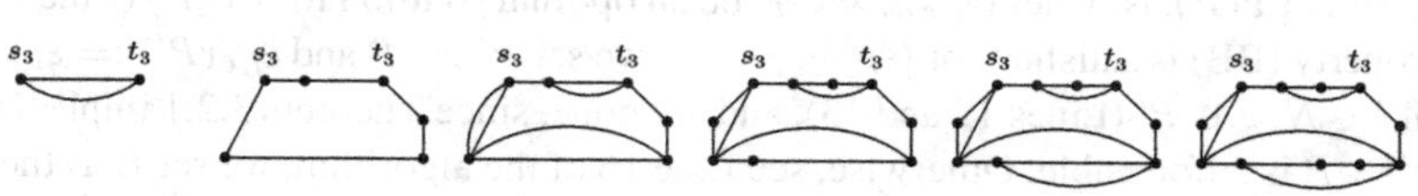

Construction of G_3 from $s_3 - t_3$ by series and parallel operations (first parallel, as well as last parallel and last subdivide operation(s) are left out).

Fig. 3.78 Constructions for the subgraphs G_i for $i \in \{1, 2, 3\}$ (cf. Figure 3.77)

The rest of this section is organized as follows. In Subsection 3.4.1, we present Algorithm n-SEPA, which is the basis of our results, and analyze it in Subsection 3.4.2. In Subsection 3.4.3, we formally state our results and present proofs which are based on the analysis of Algorithm n-SEPA. To conclude the section, we briefly discuss our results in Subsection 3.4.4, in particular the special setting of n-series-parallel graphs.

3.4.1 The Algorithm n-SePa

The Algorithm n-SEPA takes as input a network $\mathcal{N} = (G, (s_i, t_i)_{i \in N}, c, d)$, and an arbitrary strategy profile $P = (P_1, \ldots, P_n) \in \mathcal{P}$, where $(G, (s_i, t_i)_{i \in N})$ is n-series-parallel, and the edge costs $c = (c_e)_{e \in E}$ are nondecreasing and discrete-concave. The output of the algorithm is a strategy profile $P' \in \mathcal{P}$ with cost $C(P') \le C(P)$, as well as a feasible solution $(\xi_{i,e}(P'))_{i \in N, e \in P'_i}$ for LP(P') with property (BB), showing that P' is enforceable.

We now give a rather high-level description of Algorithm n-SEPA. After setting the binary variable enforceable as *false* in Line 1, we solve LP(P), see Line 4 of Algorithm n-SEPA. Note that LP(P) either is infeasible, or it has an optimal solution. Furthermore note that in the presence of delays, LP(P) can be infeasible, whereas for the special case without delays it is always feasible ($\xi_{i,e} = 0$ for all $i \in N$, $e \in P_i$ is a feasible solution of LP(P)). In fact, for the general case with delays, LP(P) is feasible if and only if $\xi_{i,e} = 0$ for all $i \in N$, $e \in P_i$ is a feasible solution. This implies, if LP(P) is infeasible, that there is a player i and an alternative path P'_i satisfying $\sum_{e \in P_i \setminus P'_i} d_{i,e} > \sum_{e \in P'_i \setminus P_i} (c_e(n_e(P) + 1) + d_{i,e})$. In Line 6 of Algorithm n-SEPA, we choose such an alternative which is *minimally violated*, meaning that some additional properties have to be satisfied (for more details, see Definition 3.4.2). Now the strategy profile P is updated by substituting strategy P_i by P'_i (Line 7). In the proof of Lemma 3.4.5 we show that this strictly decreases the overall cost of the profile P. By repeating this procedure, since the number of strategy profiles is finite, we eventually get a profile P having no larger cost than the profile we started with, and with feasible LP(P). Now let $(\xi_{i,e})_{i \in N, e \in P_i}$ be an optimal solution for LP(P) (Line 9). If property (BB) is satisfied for $(\xi_{i,e})_{i \in N, e \in P_i}$, we set $P' := P$ and $\xi_{i,e}(P') := \xi_{i,e}$ for all $i \in N$, $e \in P'_i$ (Lines 12 and 13), and are done, since Theorem 3.2.1 implies that $P' = P$ is enforceable. Otherwise, see Line 16 of the algorithm, we set $\bar{E}$ as the set of all edges which are used in P, but not completely paid according to $(\xi_{i,e})_{i \in N, e \in P_i}$, i.e., $e \in \bar{E}$ if and only if $N_e(P) \ne \emptyset$ and $\sum_{i \in N_e(P)} \xi_{i,e} < c_e(P)$. Furthermore, we set the binary variable paid as *false* (see Line 17). In Line 19, the algorithm initializes P' by P, and afterwards, it changes P' so that the edges in $\bar{E}$ are not used anymore. In the following, we describe this in some more detail. First note that, whenever a player j uses an edge $f \in \bar{E}$ in her path $P'_j = P_j$, she has an alternative path $P''_j \in \mathcal{P}_j$ with $f \notin P''_j$, for which the corresponding LP(P)-inequality is tight, i.e., $\sum_{e \in P_j \setminus P''_j} (\xi_{j,e} + d_{j,e}) = \sum_{e \in P''_j \setminus P_j} ((c_e(P'_j, P_{-j})) + d_{j,e})$. The existence of P_j'' follows from the optimality of the solution $(\xi_{i,e})_{i\in N, e\in P_i}$ for LP(P), because otherwise increasing $\xi_{j,f}$ by some small amount, while all other variables remain unchanged, yields a feasible LP(P)-solution with higher objective function value. Furthermore, if $\bar{E}_j := \bar{E} \cap P_j' \neq \emptyset$ is the set of edges in $\bar{E}$ used by player j, there is a path P_j'' which does not use any of the edges in $\bar{E}_j$, i.e., $P_j'' \cap \bar{E}_j = \emptyset$, and with tight inequality in LP(P). If P_j'' additionally satisfies some technical properties, we call P_j'' a *smallest tight alternative for* $\bar{E}_j$ (for more details, see Definition 3.4.3). In Line 21, the algorithm now updates P': For each player i who uses some edges in $\bar{E}$, we substitute P_i' with a smallest tight alternative for $\bar{E}_i$. Furthermore, the algorithm defines $(\xi_{i,e}(P'))_{i\in N, e\in P_i'}$ as follows (see Line 23): If e is an edge that player i uses in her current path P_i', but not in P_i, she now pays the complete edge cost (in the case that only she deviates), i.e., $\xi_{i,e}(P') := c_e((P_i', P_{-i}))$ for $e \in P_i' \setminus P_i$. For edges already used in P_i, the value remains unchanged, $\xi_{i,e}(P') = \xi_{i,e}$. Since we use tight alternatives, this does not change the private costs of the players, that is, $\sum_{e\in P_i'} (\xi_{i,e}(P') + d_{i,e}) = \sum_{e\in P_i} (\xi_{i,e} + d_{i,e})$ for all players i. After setting paid to *true* in Line 24, the algorithm checks for each edge e used in P' if e is not completely paid according to $(\xi_{i,e}(P'))_{i\in N, e\in P_i'}$, i.e., if $\sum_{i\in N_e(P')} \xi_{i,e}(P') < c_e(P')$ holds. In this case, we add e to $\bar{E}$ (Line 27) and set paid to *false* (Line 28). This results in another iteration of the second inner repeat-loop (Line 18–Line 29). That is, the algorithm repeats the procedure of changing P so that the edges in $\bar{E}$ are not used anymore, where $\bar{E}$ now contains all edges which are not completely paid according to $(\xi_{i,e})_{i\in N, e\in P_i}$, together with all unpaid edges according to $(\xi_{i,e}(P'))_{i\in N, e\in P_i'}$. We show in Subsection 3.4.2.1 and Subsection 3.4.2.3 that the second inner repeat-loop terminates with a cheaper strategy profile, i.e., we eventually get a strategy profile P' with $C(P') < C(P)$, and cost shares $(\xi_{i,e}(P'))_{i\in N, e\in P_i'}$ such that all used edges are completely paid. In this case, the binary variable paid is *true* at the end of the second inner repeat-loop, so the algorithm continues with Line 30. Note that there can be "overpaid" edges, i.e., with $\sum_{i\in N_e(P')} \xi_{i,e}(P') > c_e(P')$, for example if there are two players which do not use an edge e with a constant edge cost $c_e > 0$ in their paths under P, but use it in P', and therefore both pay c_e. In Line 30, the algorithm decreases the cost shares for overpaid edges arbitrarily until we reach $\sum_{i\in N_e(P')} \xi_{i,e}(P') = c_e(P')$. The algorithm now tests in Line 31 if the cost shares $(\xi_{i,e}(P'))_{i\in N, e\in P_i'}$ are a feasible solution of LP(P'). Note that for constant edge cost functions, the computed cost shares are always feasible, but for load-dependent functions, this does not hold in general (see Subsection 3.4.2.3). If the cost shares are feasible, we set enforceable to *true* and are done, since Theorem 3.2.1 implies that P' is enforceable. Otherwise, we replace P with P' and repeat the complete procedure. Note that in each iteration, in which the algorithm does not terminate, the cost of P is strictly decreased. Since the number of strategy profiles is finite, we conclude that the algorithm eventually terminates with an enforceable strategy profile (see Subsection 3.4.2.3).

We now illustrate in Figure 3.79 how a strategy profile P may be transformed during Algorithm n-SEPA, in particular during the second inner repeat-loop (Line 18–Line 29 of the algorithm). Note that we do not give specific costs or cost shares since we only want to demonstrate the possible changes on P. Assume that $LP(P)$ is feasible, and assume further, with respect to an optimal $LP(P)$-solution, that the edge e is the only edge which is not completely paid. Finally, assume that Player 1 pays a positive amount $\xi_{1,f} > 0$ for edge f. Since both players use edge e, they each have a smallest tight alternative for e. Assume that for Player 1, this alternative uses the dashed edge, whereas Player 2's smallest tight alternative for e uses the dotted path (see Figure 3.79a). In the first iteration of the second inner repeat-loop, both players substitute e by using their tight alternatives (as displayed in Figure 3.79b). But now, it may be the case that edge f is not completely paid (for example if the edge cost of f is constant): Note that Player 1 does not use edge f anymore, and thus in particular does not pay her cost share $\xi_{1,f} > 0$ for f anymore. As we show in Subsection 3.4.2.2, Player 2 then has a smallest tight alternative for f. Assuming that this alternative uses the dotted path in Figure 3.79b, we get the strategy profile given in Figure 3.79c if e and f are substituted in the second iteration of the second inner repeat-loop. Now, all edges are completely paid.

(a) At the beginning: e is not completely paid; $\bar{E} = \{e\}$.

(b) After both players substituted e: f is not completely paid; $\bar{E} = \{e, f\}$.

(c) At the end: All used edges are completely paid.

Fig. 3.79 Illustration for Algorithm n-SEPA: Used edges are thick .

Algorithm 3: n-SEPA

Input: Network $\mathcal{N} = (G, (s_i, t_i)_{i \in N}, c, d)$ with $(G, (s_i, t_i)_{i \in N})$ n-series-parallel and nondecreasing and discrete-concave edge costs $c = (c_e)_{e \in E}$; strategy profile $P = (P_1, \ldots, P_n) \in \mathcal{P}$

Output: Enforceable profile $P' \in \mathcal{P}$ with cost $C(P') \leq C(P)$; feasible solution $(\xi_{i,e}(P'))_{i \in N, e \in P'_i}$ for $LP(P')$ satisfying property (BB).

1 enforceable $\leftarrow$ *false*;
2 **repeat**
3   **repeat**
4   Solve $LP(P)$;
5   **if** $LP(P)$ *is infeasible* **then**
6   Let $i \in N$ and $P'_i \in \mathcal{P}_i$ with $\sum_{e \in P_i \setminus P'_i} d_{i,e} > \sum_{e \in P'_i \setminus P_i} (c_e((P'_i, P_{-i})) + d_{i,e})$ *minimally violated*;
7           $P \leftarrow (P'_i, P_{-i})$;
8   **else**
9   Let $(\xi_{i,e})_{i \in N, e \in P_i}$ be the computed optimal solution;
10   **until** $LP(P)$ *is feasible*
11   **if** *(BB) holds for* $(\xi_{i,e})_{i \in N, e \in P_i}$ **then**
12       $P' \leftarrow P$;
13       $\xi_{i,e}(P') \leftarrow \xi_{i,e}$ for all $i \in N, e \in P'_i$;
14   enforceable $\leftarrow$ *true*;
15   **else**
16   Let $\bar{E} \neq \emptyset$ be the set of edges which are not completely paid according to $(\xi_{i,e})_{i \in N, e \in P_i}$;
17   paid $\leftarrow$ *false*;
18   **repeat**
19          $P' \leftarrow P$;
20   **foreach** $i \in N$ *with* $\bar{E}_i := \bar{E} \cap P'_i \neq \emptyset$ **do**
21   Replace the path P'_i with a *smallest tight alternative for* $\bar{E}_i$;
22   **foreach** $i \in N$ *and* $e \in P'_i$ **do**
23             $\xi_{i,e}(P') = \begin{cases} \xi_{i,e}, & \text{for } e \in P'_i \cap P_i, \\ c_e((P'_i, P_{-i})), & \text{for } e \in P'_i \setminus P_i. \end{cases}$
24   paid $\leftarrow$ *true*;
25   **foreach** $e \in E$ *with* $N_e(P') \neq \emptyset$ **do**
26   **if** $\sum_{i \in N_e(P')} \xi_{i,e}(P') < c_e(P')$ **then**
27                $\bar{E} \leftarrow \bar{E} \cup \{e\}$;
28   paid $\leftarrow$ *false*;
29   **until** *paid* $=$ *true*
30   If there are overpaid edges, decrease the corresponding cost shares arbitrarily until all edges are exactly paid;
31   **if** $(\xi_{i,e}(P'))_{i \in N, e \in P'_i}$ *is feasible for* $LP(P')$ **then**
32   enforceable $\leftarrow$ *true*;
33   **else**
34          $P \leftarrow P'$;
35 **until** *enforceable* $=$ *true*

We now introduce some notation which is used in Algorithm n-SEPA.

Definition 3.4.2 ((minimally) violated paths). Let P be a strategy profile such that $LP(P)$ is infeasible. For $i \in N$, we call a path $P_i' \in \mathcal{P}_i$ *violated (for P)*, if

$$\sum_{e \in P_i \setminus P_i'} d_{i,e} > \sum_{e \in P_i' \setminus P_i} (c_e((P_i', P_{-i})) + d_{i,e}) \tag{3.3}$$

holds. We call P_i' *minimally violated (for P)*, if P_i' is violated for P and additionally satisfies the following properties (see Figure 3.80 for illustration, as well as our explanations thereafter):

1. $P_i \cup P_i'$ contains a unique cycle.
2. There is no violated path $P_i'' \in \mathcal{P}_i$ (for P) such that $P_i \setminus P_i'' \subsetneq P_i \setminus P_i'$.
3. For all violated paths $P_i'' \in \mathcal{P}_i$ (for P) such that $P_i \setminus P_i'' = P_i \setminus P_i'$, it holds that $\sum_{e \in P_i' \setminus P_i} (c_e((P_i', P_{-i})) + d_{i,e}) \leq \sum_{e \in P_i'' \setminus P_i} (c_e((P_i', P_{-i})) + d_{i,e})$.

Whenever it is clear from the context, we drop the specification *for P* and just speak of *(minimally) violated* paths. Note that if $LP(P)$ is infeasible, there always exists a minimally violated path (thus Line 6 of Algorithm n-SEPA can be executed). This is now explained in detail.

We now give further explanations for the conditions of Definition 3.4.2, and for the existence of (minimally) violated paths. Precisely, we argue that $LP(P)$ is infeasible if and only if there is a violated path for some player i, and that this is again equivalent to the existence of a *minimally* violated path for player i. If $LP(P)$ is infeasible, then in particular the solution $\xi_{i,e} = 0$ for all $i \in N, e \in P_i$ is infeasible. Thus, there exists a violated path $P_i' \in \mathcal{P}_i$ for some player i. On the other hand, the existence of a violated path obviously implies that $LP(P)$ is infeasible (since the variables need to be nonnegative). Next, we argue that if there is a violated path for P for some player i, there also exists a minimally violated path for i (the other direction holds trivially). Let P_i' be a violated path for P. Since $P_i' \neq P_i$ holds, $P_i \cup P_i'$ contains at least one cycle. Furthermore, each such cycle consists of a subpath p_i of P_i and a subpath p_i' of P_i', and $(P_i \setminus p_i) \cup p_i'$ is an (s_i, t_i)-path such that the union with P_i contains exactly one cycle. It is clear that among all paths that can be created like this from P_i', there is at least one *violated* path. Thus there is a violated path fulfilling the first property of minimally violated paths given in Definition 3.4.2. Assume that P_i' already satisfies this property, i.e., $P_i \cup P_i'$ contains a unique cycle consisting of

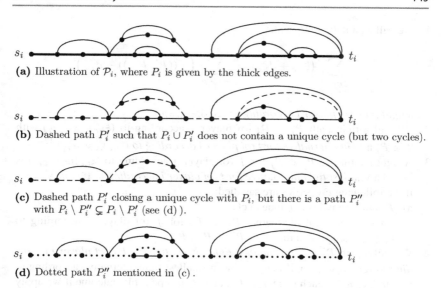

(a) Illustration of $\mathcal{P}_i$, where P_i is given by the thick edges.

(b) Dashed path P_i' such that $P_i \cup P_i'$ does not contain a unique cycle (but two cycles).

(c) Dashed path P_i' closing a unique cycle with P_i, but there is a path P_i'' with $P_i \setminus P_i'' \subsetneq P_i \setminus P_i'$ (see (d)).

(d) Dotted path P_i'' mentioned in (c).

Fig. 3.80 Illustration for (minimally) violated paths (Definition 3.4.2)

the subpaths p_i of P_i and p_i' of P_i'. Note that if player i changes her strategy from P_i to P_i', the subpath p_i is substituted by p_i'. The second property of minimally violated paths from Definition 3.4.2 means that P_i' is minimal in the sense that no other violated path substitutes a strict subset of p_i. The third property additionally requires that among all violated paths substituting exactly p_i, the path P_i' is also cost-minimal, meaning that costs plus delays for the "new" edges are minimal. Since the number of paths is finite, it is clear that the second and third property can also be satisfied, thus there exists a minimally violated path for player i, as desired.

Similar as in Section 3.3, where we analyzed network cost sharing games with two players, we again use the concept of tight paths/alternatives, meaning that the corresponding inequality (2) of LP(P) is tight (and some additional properties are fulfilled).

Definition 3.4.3 (tight paths; (smallest) tight alternatives). Let P be a strategy profile with feasible LP(P), and let $(\xi_{i,e})_{i \in N, e \in P_i}$ be an optimal solution of LP(P). Furthermore, for a player j, let $f \in P_j$, and $E_j \subseteq P_j$.

1. We call a path $P'_j \in \mathcal{P}_j$ with

$$\sum_{e \in P_j \setminus P'_j} \left(\xi_{j,e} + d_{j,e} \right) = \sum_{e \in P'_j \setminus P_j} \left(c_e((P'_j, P_{-j})) + d_{j,e} \right)$$

a *tight path (of player j according to* $(\xi_{i,e})_{i \in N, e \in P_i}$).

2. We call a tight path $P'_j \in \mathcal{P}_j$ (of player j according to $(\xi_{i,e})_{i \in N, e \in P_i}$) with $f \notin P'_j$ a *tight alternative for f (of player j according to* $(\xi_{i,e})_{i \in N, e \in P_i}$).[12]

3. A tight alternative $P'_j \in \mathcal{P}_j$ for f (of player j according to $(\xi_{i,e})_{i \in N, e \in P_i}$) is called a *smallest tight alternative for f (of player j according to* $(\xi_{i,e})_{i \in N, e \in P_i}$), if the following properties are satisfied:

   a) $P_j \cup P'_j$ contains a unique cycle.
   b) There is no tight alternative $P''_j \in \mathcal{P}_j$ for f (of player j according to $(\xi_{i,e})_{i \in N, e \in P_i}$) such that $P_j \setminus P''_j \subsetneq P_j \setminus P'_j$.

4. We call a path $P'_j \in \mathcal{P}_j$ a *smallest tight alternative for E_j (of player j according to* $(\xi_{i,e})_{i \in N, e \in P_i}$), if player j has a smallest tight alternative according to $(\xi_{i,e})_{i \in N, e \in P_i}$ for each edge $f \in E_j$, and P'_j is a possible outcome if we apply Algorithm TIGHT for P, $(\xi_{i,e})_{i \in N, e \in P_i}$, j and E_j (see page 147f. and Figure 3.81 for an illustrative example and explanations for Algorithm TIGHT).

Whenever it is clear from the context, we drop the specifications *of player j* and *according to* $(\xi_{i,e})_{i \in N, e \in P_i}$, and just speak of a *tight path*, and a *(smallest) tight alternative for f* (E_j).

Remark 3.4.4 Note that the definition of a *smallest* tight alternative is similar to the definition of a *minimally* violated path (see Definition 3.4.2): Both require that the union with P_j contains a unique cycle and that there is no other path (tight or violated, respectively) substituting a strictly smaller subset of P_j. By the same argumentation as for (minimally) violated paths, we thus get that player j has a tight alternative for f if and only if she has a *smallest* tight alternative for f. Finally note that the existence of a tight alternative for f is guaranteed if edge f is not completely paid according to $(\xi_{i,e})_{i \in N, e \in P_i}$: Otherwise, increasing $\xi_{j,f}$ by some small amount, while all other variables remain unchanged, yields a feasible LP(P)-solution with higher objective, contradicting the optimality of $(\xi_{i,e})_{i \in N, e \in P_i}$.

---

[12]Note that in Section 3.3, the term *tight alternative* is mainly used for a certain subpath of P'_j (instead of the complete path P'_j), see Notation (N3).

Algorithm 4: TIGHT

Input: Network $\mathcal{N} = (G, (s_i, t_i)_{i \in N}, c, d)$ with $(G, (s_i, t_i)_{i \in N})$
n-series-parallel; strategy profile $P = (P_1, \ldots, P_n) \in \mathcal{P}$ with
$(\xi_{i,e})_{i \in N, e \in P_i}$ optimal solution of LP(P); $j \in N$ and $E_j \subseteq P_j$ such
that j has a tight alternative for each edge $f \in E_j$.

Output: Tight path $P_j' \in \mathcal{P}_j$ with $P_j' \cap E_j = \emptyset$.

1 **foreach** $f \in E_j$ **do**

2 $\quad$ Let P_j^f be a smallest tight alternative for f;

3 $\quad$ $A_{j,f} \leftarrow P_j^f \setminus P_j$;

4 $\quad$ $S_{j,f} \leftarrow P_j \setminus P_j^f$;

5 $\quad$ $b_f \leftarrow 0$;

6 $H \leftarrow \{P_j^f : f \in E_j\}$;

7 **foreach** $P_j^f \in H$ **do**

8 $\quad$ **if** $b_f = 0$ **then**

9 $\quad\quad$ **foreach** $P_j^{f'} \in H$ with $P_j^{f'} \neq P_j^f, b_{f'} = 0$ and $S_{j,f'} \subseteq S_{j,f}$ **do**

10 $\quad\quad\quad$ $b_{f'} \leftarrow 1$;

11 $P_j' \leftarrow P_j$;

12 **foreach** $P_j^f \in H$ **do**

13 $\quad$ **if** $b_f = 0$ **then**

14 $\quad\quad$ $P_j' \leftarrow P_j' \setminus S_{j,f} \cup A_{j,f}$.

Figure 3.81 illustrates the notions introduced in Definition 3.4.3 and Algorithm TIGHT. Assume that $E_j = \{f, g, h\}$, and that the subsets $A_{j,f}, A_{j,g}, A_{j,h}$ and $S_{j,f}, S_{j,g}, S_{j,h}$ corresponding to the tight alternatives P_j^f, P_j^g and P_j^h are as displayed in (b) and (c) of Figure 3.81. Furthermore assume that the algorithm always considers H in the order P_j^f, P_j^h, P_j^g. In the first iteration of Line 7–Line 10 (where we consider P_j^f), nothing happens, since neither $S_{j,h} \subseteq S_{j,f}$, nor $S_{j,g} \subseteq S_{j,f}$ holds. In the second iteration (P_j^h), the variable b_g is set to 1 since $S_{j,g} \subseteq S_{j,h}$. In the last iteration (P_j^g), nothing happens since $b_g = 1$ holds. In the first iteration of Line 12–Line 14 (where we consider P_j^f), the path $P_j = P_j'$ is replaced by $P_j^f = P_j' \setminus S_{j,f} \cup A_{j,f}$ (see (d) of Figure 3.81). Note that replacing P_j with a smallest tight alternative P_j^f means that the set of edges $S_{j,f} \subseteq P_j$ is substituted by the set $A_{j,f} \subseteq P_j^f$. Therefore we also denote $A_{j,f}$ as a *smallest tight alternative for* f (instead of P_j^f); and $S_{j,f}$ is denoted as the *edges which are substituted by* $A_{j,f}$. In the second iteration of Line 12–Line 14 ($P_{j,h}$), P_j' is replaced by $P_j' \setminus S_{j,h} \cup A_{j,h}$ (see (e) of Figure 3.81). In the last iteration ($P_{j,g}$), the path P_j' remains unchanged since $b_g = 1$.

(a) Illustration of $\mathcal{P}_j$, where P_j is given by the thick edges, and $E_j = \{f, g, h\}$.

(b) $A_{j,f}$ (thick), $A_{j,g}$ (dashed) and $A_{j,h}$ (dotted) as defined in Algorithm TIGHT, corresponding to the smallest tight alternatives P_j^f, P_j^g and P_j^h.

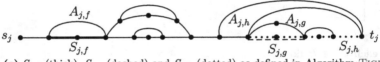

(c) $S_{j,f}$ (thick), $S_{j,g}$ (dashed) and $S_{j,h}$ (dotted) as defined in Algorithm TIGHT.

(d) Path P_j' (thick) after the first iteration of Line 12 - Line 14 of Algorithm TIGHT ($S_{j,f}$ is substituted by $A_{j,f}$).

(e) Path P_j' (thick) computed by Algorithm TIGHT, i.e., a smallest tight alternative for E_j.

Fig. 3.81 Illustration of (smallest) tight alternatives (Definition 3.4.3 and Algorithm TIGHT)

Algorithm TIGHT computes a "suitable combination" of the smallest tight alternatives for E_j, such that the result is a tight (s_j, t_j)-path containing no edges of E_j.

Note that if there are two edges $f, g \in E_j$ with tight alternatives P_j^f and P_j^g such that $S_{j,f} \subseteq S_{j,g}$ holds, i.e., P_j^f substitutes a subset of the edges which are substituted by P_j^g, we "do not need" $P_{j,f}$, since $P_{j,g}$ also substitutes f. The algorithm thus chooses a subset of the smallest tight alternatives having the property that any two alternatives P_j^f and P_j^g neither fulfill $S_{j,f} \subseteq S_{j,g}$, nor $S_{j,g} \subseteq S_{j,f}$. Due to the series-parallel structure, this already implies $S_{j,f} \cap S_{j,g} = \emptyset$. Furthermore, the series-parallel structure also yields that $A_{j,f}$ and $A_{j,g}$ are internal node-disjoint.[13] Therefore, if P_j' results from P_j by substituting $S_{j,f}$ with $A_{j,f}$ for all of the chosen smallest tight alternatives P_j^f, then P_j' is a (s_j, t_j)-path which does not use any edges of E_j. Furthermore, P_j' is also tight, since the tightness of P_j^f implies (due to the definitions of $A_{j,f}$ and $S_{j,f}$)

$$\sum_{e \in S_{j,f}} \left(\xi_{j,e} + d_{j,e} \right) = \sum_{e \in A_{j,f}} \left(c_e((P_j^f, P_{-j})) + d_{j,e} \right),$$

and thus

$$
\begin{aligned}
\sum_{e \in P_j \setminus P_j'} \left(\xi_{j,e} + d_{j,e} \right) &= \sum_{P_j^f \in H : b_f = 0} \sum_{e \in S_{j,f}} \left(\xi_{j,e} + d_{j,e} \right) \\
&= \sum_{P_j^f \in H : b_f = 0} \sum_{e \in A_{j,f}} \left(c_e((P_j^f, P_{-j})) + d_{j,e} \right) \\
&= \sum_{e \in P_j' \setminus P_j} \left(c_e((P_j', P_{-j})) + d_{j,e} \right)
\end{aligned}
$$

holds. Altogether, we observe that Algorithm TIGHT is correct, i.e., the computed path P_j' is a tight (s_j, t_j)-path and does not use any edges of E_j. It is furthermore clear that Algorithm TIGHT terminates.

---

[13] Note that the mentioned two properties do not hold in general graphs. Consider, for example, NBC2 in Definition 3.3.14: The two alternatives q_2 and q_3 are neither internal node-disjoint, nor do they substitute disjoint edge sets.

3.4.2 Analysis of Algorithm n-SePa

In this subsection, we analyze Algorithm n-SEPA, in particular we show that the algorithm is correct (Subsection 3.4.2.1), well-defined (Subsection 3.4.2.2) and terminates (Subsection 3.4.2.3). In Subsection 3.4.2.3, we furthermore analyze the running time of the algorithm. These results are then used in Subsection 3.4.3 to prove our main results.

3.4.2.1 Correctness of Algorithm n-SePa

The aim of this subsection is to show the following: If Algorithm n-SEPA terminates with strategy profile P' and cost shares $(\xi_{i,e}(P'))_{i \in N, e \in P_i'}$, then the cost of P' is at most the cost of the input profile P, and the computed cost shares for P' are a feasible solution for LP(P') satisfying property (BB).

We need the following lemma, which shows that the update of P in Line 7 of Algorithm n-SEPA decreases the cost of the strategy profile.

Lemma 3.4.5 *Let P be a strategy profile such that LP(P) is infeasible, $P_i' \in \mathcal{P}_i$ a violated path (for some player i), and $P' := (P_i', P_{-i})$. Then $C(P') < C(P)$.*

Proof For a strategy profile P'' and an edge e, define for convenience $c_e(P'') := 0$ if $N_e(P'') = \emptyset$, i.e., edge e is not used in P''. We then get

$$
\begin{aligned}
C(P) - C(P') = {} & \sum_{e \in P_i \setminus P_i'} \left(c_e(P) + \sum_{j \in N_e(P)} d_{j,e} \right) + \sum_{e \in P_i' \setminus P_i} \left(c_e(P) + \sum_{j \in N_e(P)} d_{j,e} \right) \\
& - \sum_{e \in P_i \setminus P_i'} \left(c_e(P') + \sum_{j \in N_e(P')} d_{j,e} \right) - \sum_{e \in P_i' \setminus P_i} \left(c_e(P') + \sum_{j \in N_e(P')} d_{j,e} \right) \\
= {} & \sum_{e \in P_i \setminus P_i'} \left(c_e(P) - c_e(P') \right) + \sum_{e \in P_i \setminus P_i'} d_{i,e} - \sum_{e \in P_i' \setminus P_i} \left(c_e(P') + d_{i,e} \right) \\
& + \sum_{e \in P_i' \setminus P_i} c_e(P).
\end{aligned}
$$

Since edge costs are nondecreasing, $c_e(P) \geq c_e(P')$ holds for edges $e \in P_i \setminus P_i'$. Furthermore we get, by definition of a violated path, that $\sum_{e \in P_i \setminus P_i'} d_{i,e} > \sum_{e \in P_i' \setminus P_i} (c_e(P') + d_{i,e})$. Finally, $c_e(P) \geq 0$ holds, and altogether this shows $C(P) - C(P') > 0$. $\square$

Note that in Lemma 3.4.5, P_i' does not need to be *minimally* violated, but we use the minimality to show polynomial running time of Algorithm *n*-SEPA for a special case (see Theorem 3.4.20).

In the next lemma, we show that if the variable paid is *true* at the end of an iteration of the second inner repeat-loop (Line 18–Line 29), then the update of P' in Line 20–Line 21 of this iteration strictly decreased the cost of P'.

Lemma 3.4.6   *Let P be a strategy profile which is not enforceable, but with feasible LP(P), and let $(\xi_{i,e})_{i\in N, e\in E}$ be an optimal solution of LP(P). Assume that we apply Line 16–Line 29 of Algorithm n-SEPA for P and $(\xi_{i,e})_{i\in N, e\in E}$, and that we terminate with P' and $(\xi_{i,e}(P'))_{i\in N, e\in P_i'}$. Then $C(P') < C(P)$ holds.*

Proof Since P is not enforceable, property (BB) is not satisfied for $(\xi_{i,e})_{i\in N, e\in E}$. Therefore,

$$
\begin{aligned}
C(P) &= \sum_{\substack{e\in E:\\ N_e(P)\neq\emptyset}} c_e(P) + \sum_{i\in N}\sum_{e\in P_i} d_{i,e} \\
&> \sum_{\substack{e\in E:\\ N_e(P)\neq\emptyset}} \sum_{i\in N_e(P)} \xi_{i,e} + \sum_{i\in N}\sum_{e\in P_i} d_{i,e} \\
&= \sum_{i\in N}\sum_{e\in P_i} \left(\xi_{i,e} + d_{i,e}\right)
\end{aligned}
$$

holds. On the other hand,

$$
\sum_{i\in N_e(P')} \xi_{i,e}(P') \geq c_e(P')
$$

is fulfilled for all edges e with $N_e(P') \neq \emptyset$ (paid is *true*). Thus we get

$$
\begin{aligned}
C(P') &= \sum_{\substack{e\in E:\\ N_e(P')\neq\emptyset}} c_e(P') + \sum_{i\in N}\sum_{e\in P_i'} d_{i,e} \\
&\leq \sum_{\substack{e\in E:\\ N_e(P')\neq\emptyset}} \sum_{i\in N_e(P')} \xi_{i,e}(P') + \sum_{i\in N}\sum_{e\in P_i'} d_{i,e} \\
&= \sum_{i\in N}\sum_{e\in P_i'} \left(\xi_{i,e}(P') + d_{i,e}\right).
\end{aligned}
$$

Finally, for each player $j \in N$, the private cost according to P' and $(\xi_{i,e}(P'))_{i \in N, e \in P'_i}$ equals the private cost according to P and $(\xi_{i,e})_{i \in N, e \in P_i}$, i.e.,

$$\sum_{e \in P'_j} \left(\xi_{j,e}(P') + d_{j,e} \right) = \sum_{e \in P_j} \left(\xi_{j,e} + d_{j,e} \right)$$

holds: Due to the correctness of Algorithm TIGHT, the path P'_j is tight for each player j, that is,

$$\sum_{e \in P_j \setminus P'_j} \left(\xi_{j,e} + d_{j,e} \right) = \sum_{e \in P'_j \setminus P_j} \left(c_e((P'_j, P_{-j})) + d_{j,e} \right)$$

holds. Furthermore, the definition of $(\xi_{i,e}(P'))_{i \in N, e \in P'_i}$ in Line 23 of Algorithm n-SEPA yields $\xi_{j,e}(P') = \xi_{j,e}$ for all $e \in P'_j \cap P_j$, and $\xi_{j,e}(P') = c_e((P'_j, P_{-j}))$ for all $e \in P'_j \setminus P_j$. Putting everything together shows

$$\sum_{e \in P'_j} \left(\xi_{j,e}(P') + d_{j,e} \right) = \sum_{e \in P'_j \cap P_j} \left(\xi_{j,e}(P') + d_{j,e} \right) + \sum_{e \in P'_j \setminus P_j} \left(\xi_{j,e}(P') + d_{j,e} \right)$$

$$= \sum_{e \in P'_j \cap P_j} \left(\xi_{j,e} + d_{j,e} \right) + \sum_{e \in P'_j \setminus P_j} \left(c_e((P'_j, P_{-j})) + d_{j,e} \right)$$

$$= \sum_{e \in P_j \cap P'_j} \left(\xi_{j,e} + d_{j,e} \right) + \sum_{e \in P_j \setminus P'_j} \left(\xi_{j,e} + d_{j,e} \right)$$

$$= \sum_{e \in P_j} \left(\xi_{j,e} + d_{j,e} \right).$$

Using all of the above then shows

$$C(P) > \sum_{i \in N} \sum_{e \in P_i} \left(\xi_{i,e} + d_{i,e} \right) = \sum_{i \in N} \sum_{e \in P'_i} \left(\xi_{i,e}(P') + d_{i,e} \right) \geq C(P'),$$

completing the proof of Lemma 3.4.6. $\square$

We are now able to show that Algorithm n-SEPA is correct:

Proposition 3.4.7 *Assume that Algorithm n-SEPA terminates with strategy profile* P' *and cost shares* $(\xi_{i,e}(P'))_{i \in N, e \in P'_i}$. *Then* $C(P') \leq C(P)$ *holds, where* P *is* *the input profile, and* $(\xi_{i,e}(P'))_{i\in N, e\in P'_i}$ *is a feasible solution for LP(P') fulfilling property* (BB) *(thus, P' is enforceable). Furthermore, if P is enforceable, P' = P holds, and if P is not enforceable, we get* $C(P') < C(P)$.

Proof Note that Algorithm *n*-SEPA terminates after the binary variable enforceable is set to *true*. This happens either in Line 14 or in Line 32 of the algorithm. In both cases it is clear that $(\xi_{i,e}(P'))_{i\in N, e\in P'_i}$ is a feasible solution for LP(P') fulfilling property (BB).

We now turn to the cost of the output profile. First, note that whenever P is updated (see Line 7 and Line 34), this strictly decreases the cost of P (due to Lemma 3.4.5 and Lemma 3.4.6). This shows $C(\bar{P}) \leq C(P)$, where $\bar{P}$ denotes the assignment of the variable P at the time Algorithm *n*-SEPA terminates. We now distinguish between the two cases that the variable enforceable is set to *true* in Line 14 or in Line 32, and show that $C(P') \leq C(\bar{P})$ holds for both cases. Together with $C(\bar{P}) \leq C(P)$, we get $C(P') \leq C(P)$. If enforceable is set to *true* in Line 14, $P' = \bar{P}$ holds, which obviously implies $C(P') = C(\bar{P})$. Otherwise (enforceable is set to *true* in Line 32), we get $C(P') < C(\bar{P})$ from Lemma 3.4.6.

If P is enforceable, it is clear that $P' = P$ holds. To complete the proof, it remains to show $C(P') < C(P)$ if P is not enforceable. From our observations above, we get that $C(P') < C(P)$ holds if the variable P is updated at least once ($C(P') \leq C(\bar{P}) < C(P)$ holds), or if enforceable is set to *true* in Line 32 ($C(P') < C(\bar{P}) \leq C(P)$ holds). If none of the above holds, i.e., the variable P is not updated during the course of the algorithm, and enforceable is set to *true* in Line 14, we conclude $P' = \bar{P} = P$, which shows that P is enforceable (since P' is enforceable). This completes the proof: If P is not enforceable, at least one of the above holds, and this implies $C(P') < C(P)$. □

3.4.2.2 Well-Definedness of Algorithm *n*-SePa

The aim of this subsection is to show that Algorithm *n*-SEPA is well-defined, that is, each step can be executed. For all steps except the ones described in Line 6 and Line 21, this clearly holds. On page 144, we argued that a minimally violated path exists if LP(P) is infeasible, thus Line 6 can also be executed. It remains to show that the step described in Line 21 is well-defined:

Proposition 3.4.8 *In Line 21 of Algorithm n-*SEPA, *a smallest tight alternative for* $\bar{E}_i$ *always exists.*

Proof Recall that a player *i* has a *smallest* tight alternative for an edge *e* if and only if player *i* has *any* tight alternative for *e*. Furthermore, player *i* has a smallest tight alternative for a subset E' of the edges if she has smallest tight alternatives for all edges in E'. To prove Proposition 3.4.8, it is thus sufficient to show, for each edge $e \in \bar{E}_i$, that player i has a tight alternative for e.

To this end, consider an arbitrary, but fixed iteration of the outer repeat-loop with the property that (BB) does not hold for $(\xi_{i,e})_{i \in N, e \in E}$ (thus, we reach Line 21 in this iteration). For this fixed iteration, call each iteration of the second inner repeat-loop (Line 18–Line 29) a *phase of deviation*. Furthermore, $\bar{E}^k$ denotes the set of edges added to $\bar{E}$ in Line 27 of phase $k \geq 1$, and $\bar{E}^0$ are the edges which are added in *phase 0*, meaning Line 16 of the algorithm. We now show that for all $k \geq 0$, each player $j \in N$ with $\bar{E}^k_j := \bar{E}^k \cap P_j \neq \emptyset$ has a tight alternative for each edge $f \in \bar{E}^k_j$. This shows Proposition 3.4.8.

The proof proceeds by induction on k. For the base case, i.e., $k = 0$, the existence of the smallest tight alternatives follows from the optimality of $(\xi_{i,e})_{i \in N, e \in P_i}$ for LP(P) and the fact that $\bar{E}^0$ consists of the edges which are not completely paid according to $(\xi_{i,e})_{i \in N, e \in P_i}$: If there is a player j and an edge $f \in \bar{E}^0_j$ such that player j has no tight alternative for f, increasing $\xi_{j,f}$ by some small amount, while all other variables remain unchanged, yields a feasible solution for LP(P) with higher objective function value than $(\xi_{i,e})_{i \in N, e \in P_i}$; contradiction.

For the induction step, assume that the tight alternatives exist for all edges added in phases $< k$, for $k \geq 1$. Furthermore assume, by contradiction, that there is a player j and an edge $f \in \bar{E}^k_j$ such that j has no tight alternative for f. As in the base case, we deduce a contradiction to the optimality of the solution $(\xi_{i,e})_{i \in N, e \in P_i}$ for LP(P):

Let P' be the strategy profile, and let $(\xi_{i,e}(P'))_{i \in N, e \in P'_i}$ be the cost shares after the updates in Lines 20–23 of the kth phase. In Lemma 3.4.9, we show the existence of a player $\ell \in N_f(P) \setminus N_f(P')$ with $\xi_{\ell,f} > 0$. Using the notation introduced in Algorithm TIGHT, the path P'_ℓ is a combination of P_ℓ and some smallest tight alternatives $A_{\ell,g}$ for edges $g \in \bar{E}_\ell := \cup_{\bar{k} < k} \bar{E}^{\bar{k}}_\ell$. Since $f \notin P'_\ell$, but also $f \notin \bar{E}_\ell$, there has to be an edge $g \neq f$ with $g \in \bar{E}_\ell$ such that the smallest tight alternative $A_{\ell,g} \subseteq P'_\ell$ also substitutes f, i.e., $f \in S_{\ell,g}$ (see Figure 3.82 for illustration). This implies that *all* tight alternatives of player ℓ for g substitute f (due to the series-parallel structure and that $A_{\ell,g}$ is a *smallest* tight alternative for g). Therefore, if we decrease $\xi_{\ell,f}$ (by some small amount), player ℓ does not have a tight alternative for g anymore. If g is not completely paid according to $(\xi_{i,e})_{i \in N, e \in P_i}$, the following changes of $(\xi_{i,e})_{i \in N, e \in P_i}$ (for suitably small $\varepsilon > 0$) then lead to a feasible solution for LP(P) with higher objective function value than $(\xi_{i,e})_{i \in N, e \in P_i}$; contradiction: Increase $\xi_{j,f}$ by ε, decrease $\xi_{\ell,f}$ by ε, and increase $\xi_{\ell,g}$ by ε.

Otherwise, let $0 < k' < k$ be the phase in which g is added to $\bar{E}$. We can repeat the argumentation of the former paragraph for (ℓ, g, k') instead of (j, f, k), and eventually this leads to a sequence $(p_z, e_z, d_z)_{z=1,\dots,Z}$ of players p_z, edges e_z and phases of deviation d_z having the following properties:[14] For all $z = 1, \dots, Z$, edge e_z (used by player p_z) is added to $\bar{E}$ in phase d_z. For $z = 2, \dots, Z$, player p_z substitutes e_{z-1} in phase d_{z-1} by using a smallest tight alternative for e_z, and $\xi_{p_z, e_{z-1}} > 0$. Finally, the phases are decreasing, i.e., $d_{z-1} > d_z$ for all $z = 2, \dots, Z$, and $d_Z = 0$. We can now change $(\xi_{i,e})_{i \in N, e \in P_i}$ as follows: First, increase $\xi_{p_1, e_1} = \xi_{j,f}$ by ε. Then, for all $z = 2, \dots, Z$, decrease $\xi_{p_z, e_{z-1}}$ by ε and increase ξ_{p_z, e_z} by ε. By suitably small $\varepsilon > 0$, this yields a feasible solution for LP(P) with higher objective function value than $(\xi_{i,e})_{i \in N, e \in P_i}$; thus contradicting the optimality of $(\xi_{i,e})_{i \in N, e \in P_i}$, and completing the proof. □

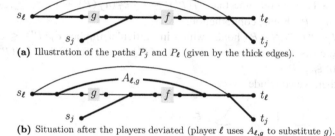

(a) Illustration of the paths P_j and P_ℓ (given by the thick edges).

(b) Situation after the players deviated (player ℓ uses $A_{\ell,g}$ to substitute g).

Fig. 3.82 Illustration for the proof that tight alternatives exist

It remains to show the following lemma, which is used in the proof of Proposition 3.4.8.

Lemma 3.4.9 *In the induction step in the proof of Proposition 3.4.8, there exists a player $\ell \in N_f(P) \setminus N_f(P')$ with $\xi_{\ell,f} > 0$.*

Proof Since f is added to $\bar{E}$ in phase $k \geq 1$, we get that $\sum_{i \in N_f(P)} \xi_{i,f} = c_f(P)$ and $\sum_{i \in N_f(P')} \xi_{i,f}(P') < c_f(P')$ holds. In particular, the strict inequality implies $N_f(P') \subseteq N_f(P)$:

---

[14]Note that $(p_1, e_1, d_1) = (j, f, k)$ and $(p_2, e_2, d_2) = (\ell, g, k')$.

Assume, by contradiction, that $|N_f(P') \setminus N_f(P)| =: r \geq 1$. Using that edge costs are nonnegative and discrete-concave yields

$$c_f(n_f(P)+1) \geq c_f(n_f(P)+1) - c_f(n_f(P)) \geq c_f(n_f(P)+i) - c_f(n_f(P)+i-1)$$

for each $i \in \{2, \ldots, r\}$. Using this together with the fact that edge costs are nondecreasing, we get

$$
\begin{aligned}
c_f(P') = c_f(n_f(P')) &= c_f(|N_f(P') \cap N_f(P)| + |N_f(P') \setminus N_f(P)|) \\
&\leq c_f(n_f(P) + r) = c_f(n_f(P) + 1) + \sum_{i=2}^{r} \big(c_f(n_f(P)+i) - c_f(n_f(P)+i-1)\big) \\
&\leq c_f(n_f(P) + 1) + (r-1) \cdot c_f(n_f(P) + 1) \\
&= r \cdot c_f(n_f(P) + 1).
\end{aligned}
$$

Since each player $i \in N_f(P') \setminus N_f(P)$ pays $\xi_{i,f}(P') = c_f((P_i', P_{-i})) = c_f(n_f(P) + 1)$, we conclude that $c_f(P') \leq r \cdot c_f(n_f(P) + 1) \leq \sum_{i \in N_f(P')} \xi_{i,f}(P') < c_f(P')$ holds; contradiction.

Thus $N_f(P') \subseteq N_f(P)$ holds, which in particular implies $c_f(P') \leq c_f(P)$ (since edge costs are nondecreasing) and $\xi_{i,f}(P') = \xi_{i,f}$ for all $i \in N_f(P')$ (by definition of $\xi_{i,f}(P')$).

Altogether, we conclude

$$
\begin{aligned}
\sum_{i \in N_f(P')} \xi_{i,f} &= \sum_{i \in N_f(P')} \xi_{i,f}(P') \\
&< c_f(P') \leq c_f(P) = \sum_{i \in N_f(P)} \xi_{i,f} \\
&= \sum_{i \in N_f(P) \cap N_f(P')} \xi_{i,f} + \sum_{i \in N_f(P) \setminus N_f(P')} \xi_{i,f} \\
&= \sum_{i \in N_f(P')} \xi_{i,f} + \sum_{i \in N_f(P) \setminus N_f(P')} \xi_{i,f},
\end{aligned}
$$

which shows that a player $\ell \in N_f(P) \setminus N_f(P')$ with $\xi_{\ell,f} > 0$ exists. $\square$

3.4.2.3 Termination and Running Time of Algorithm n-SePa

In this subsection, we first show that Algorithm n-SEPA terminates, and then analyze its running time. According to the running time analysis, we want to emphasize that we were only interested in the question whether the running time is polynomial or not. Consequently, we mostly do not give specific bounds.

Proposition 3.4.10 *Algorithm n-SEPA terminates.*

Proof We need to show that each of the three repeat-loops terminates. In the first inner repeat-loop (Line 3–Line 10), we solve LP(P) until we get a strategy profile P with feasible LP(P). Since the cost of the strategy profile strictly decreases in each iteration in which LP(P) is not yet feasible (see Lemma 3.4.5), this terminates (since the number of strategy profiles is finite). Consider the second inner repeat-loop (Line 18–Line 29). In each iteration ending with paid $=$ *false*, at least one edge of $\cup_{i \in N} P_i$ is added to $\bar{E}$, since edges which are not used in P are completely paid (compare the proof of Lemma 3.4.9). Therefore, the number of iterations of the second inner repeat-loop is $O(|E|)$ and in particular, the second inner repeat-loop terminates. Finally, in each iteration of the outer repeat-loop in which we do not terminate, the cost of the variable P is strictly decreased (due to Lemma 3.4.5 and Lemma 3.4.6). Since the number of strategy profiles is finite, this shows that Algorithm n-SEPA terminates after finitely many iterations of the outer repeat-loop. □

We now further analyze the running time of Algorithm n-SEPA. Unfortunately, we do not know in general whether the number of iterations of the first inner repeat loop, or of the outer repeat-loop, is polynomially bounded. Only for the special case of constant edge cost functions, we are able to derive such bounds (see Proposition 3.4.15 and Proposition 3.4.16). Besides the number of iterations of the three repeat-loops, the main steps influencing the running time of Algorithm n-SEPA are solving LP(P), computing minimally violated paths and computing smallest tight alternatives. Each of these steps requires analyzing certain properties for *all* (s_i, t_i)-paths of player $i \in N$, which can be exponentially many paths. In the following, we show how to obtain a set $\mathcal{A}_i$ for each player i, such that $|\mathcal{A}_i| = O(|E|)$, and if the desired properties hold for $\mathcal{A}_i$, they also hold for all paths in $\mathcal{P}_i$. Furthermore, we show that $\mathcal{A}_i$ can be computed in polynomial time.

To this end, recall that the graph G_i (induced by $\mathcal{P}_i$) essentially looks like displayed in Figure 3.83a, and we can assume w.l.o.g. that P_i is given by the thick edges. An arbitrary (s_i, t_i)-path $P_i' \in \mathcal{P}_i$ consists of subpaths of P_i together with some of the "arcs". We call these arcs *alternatives (according to P_i)*, and formally, an alternative is a path A_i which connects two nodes of P_i, but otherwise is node- and edge-disjoint with P_i. The subpath of P_i with the same endnodes as A_i is denoted by S_{A_i}, and we say that this subpath is *substituted by A_i* (see Figure 3.83b for illustration). Note that a very similar notation is used in the context of Algorithm TIGHT. Furthermore note that there can be different alternatives which substitute the same subpath of P_i (in Figure 3.83b, this holds for example for the two arcs on the left which both substitute the second and third edge of P_i). We call an alternative A_i a *cheapest alternative for S_{A_i} (according to P)*, if for all alternatives A_i' with $S_{A_i'} = S_{A_i}$, alternative A_i is "cheaper" than A_i', that is,

$$\sum_{e \in A_i} \big(c_e(n_e(P) + 1) + d_{i,e}\big) \leq \sum_{e \in A_i'} \big(c_e(n_e(P) + 1) + d_{i,e}\big)$$

holds. Finally, we call a set $\mathcal{A}_i$ a *complete representative set of cheapest alternatives (according to P)*, if it consists of exactly one cheapest alternative (according to P) for each subpath of P_i which is substituted by any alternative. Figure 3.83c shows an example for such a set. Note that due to the series-parallel structure of G_i, the paths in $\mathcal{A}_i$ are pairwise edge-disjoint, which shows $|\mathcal{A}_i| = O(|E|)$.

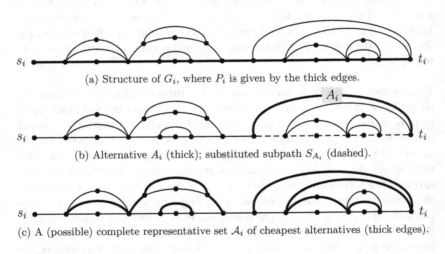

(a) Structure of G_i, where P_i is given by the thick edges.

(b) Alternative A_i (thick); substituted subpath S_{A_i} (dashed).

(c) A (possible) complete representative set $\mathcal{A}_i$ of cheapest alternatives (thick edges).

Fig. 3.83 Illustration for (cheapest) alternatives

Using breadth first search (BFS) and shortest paths, it is quite straighforward to compute a complete representative set of cheapest alternatives, see Algorithm ALTERNATIVES. In this algorithm, we consider the nodes of P_i ordered from s_i to t_i. For a vertex u chosen in Line 3, we compute in the repeat-loop, for each node v after u (in P_i) such that the subpath of P_i with endnodes u and v is substituted by any alternative, a cheapest such alternative. This obviously yields a complete representative set of cheapest alternatives. Since there are $O(|V|)$ iterations of the repeat-loop for a fixed vertex u, and executing a BFS and computing a shortest path is possible in polynomial time, we conclude that Algorithm ALTERNATIVES can be executed in polynomial time.

Algorithm 5: ALTERNATIVES

Input: Network $\mathcal{N} = (G, (s_i, t_i)_{i \in N}, c, d)$ with $(G, (s_i, t_i)_{i \in N})$ n-series-parallel; strategy profile $P = (P_1, \ldots, P_n) \in \mathcal{P}$, player $i \in N$.
Output: Complete representative set $\mathcal{A}_i$ of cheapest alternatives according to P.

1 Delete from G all edges of P_i;
2 Define $\tilde{c}_e := c_e(n_e(P) + 1) + d_{i,e}$ for all edges e in G;
3 **foreach** *node u in P_i, ordered from s_i to t_i* **do**
4   Mark all nodes of P_i which are not yet deleted from G as *not visited*;
5   **repeat**
6   Let G' be the graph resulting from G by deleting all *visited* nodes of P_i;
7   Starting in u, execute a BFS in G', and stop immediately if *not visited* node of P_i is found;
8   **if** *not visited node v of P_i is found* **then**
9   Let G'' be the graph resulting from G by deleting all nodes of P_i except u and v;
10   With respect to $\tilde{c}$, compute a shortest (u, v)-path A_i in G'';
11   Insert A_i in $\mathcal{A}_i$;
12   Delete from G all nodes (except u, v) and edges of A_i;
13   Mark v as *visited*;
14   **else**
15   Delete from G all nodes found during the last BFS (in particular u);
16   **until** *u has been deleted*

Using complete representative sets of cheapest alternatives, we can now derive the following about the complexity of solving LP(P), and computing minimally violated paths:

Lemma 3.4.11 *Let $\mathcal{N} = (G, (s_i, t_i)_{i \in N}, c, d)$ be a network, where $(G, (s_i, t_i)_{i \in N})$ is n-series-parallel, and let P be a strategy profile. Then we can decide in polynomial time whether LP(P) is infeasible. Furthermore, we can compute a minimally violated path, if LP(P) infeasible, and an optimal solution for LP(P), if LP(P) feasible, in polynomial time.*

Proof For each player i, let $\mathcal{A}_i$ be a complete representative set of cheapest alternatives according to P. It is clear that

$$\sum_{e\in P_i\setminus P_i'} d_{i,e} \leq \sum_{e\in P_i'\setminus P_i} \left(c_e(n_e(P)+1)+d_{i,e}\right) \quad \forall i \in N \; \forall P_i' \in \mathcal{P}_i \qquad (3.4)$$

holds if and only if

$$\sum_{e\in S_{A_i}} d_{i,e} \leq \sum_{e\in A_i} \left(c_e(n_e(P)+1)+d_{i,e}\right) \quad \forall i \in N \; \forall A_i \in \mathcal{A}_i \qquad (3.5)$$

holds. Since (3.4) is equivalent to the feasibility of LP(P), and (3.5) can be tested in polynomial time, we can decide in polynomial time whether LP(P) is feasible or not. Furthermore, if LP(P) is infeasible, we can compute a minimally violated path in polynomial time: Take a player i having a *violated alternative* A_i, i.e., with $\sum_{e\in S_{A_i}} d_{i,e} > \sum_{e\in A_i} \left(c_e(n_e(P)+1)+d_{i,e}\right)$. Among all violated alternatives for player i, choose A_i such that there is no other violated alternative A_i' with $S_{A_i'} \subsetneq S_{A_i}$. Then, $P_i' := P_i \setminus S_{A_i} \cup A_i$ is a minimally violated path for player i.

To complete the proof, assume that LP(P) is feasible. Let LP(P)$'$ be the linear program resulting from LP(P) by replacing property (2) with

$$\sum_{e\in S_{A_i}} \left(\xi_{i,e}+d_{i,e}\right) \leq \sum_{e\in A_i} \left(c_e(n_e(P)+1)+d_{i,e}\right) \quad \forall i \in N \; \forall A_i \in \mathcal{A}_i.$$

It is clear that $(\xi_{i,e})_{i\in N, e\in P_i}$ is optimal for LP(P) if and only if it is optimal for LP(P)$'$. Furthermore, since LP(P)$'$ has polynomially many variables and constraints, we can compute an optimal solution for LP(P)$'$ in polynomial time [77]. Thus, we can compute an optimal solution for LP(P) in polynomial time. $\square$

Using Lemma 3.4.11, we conclude the following about the running time of the first inner repeat-loop.

Proposition 3.4.12 *A single iteration of the first inner repeat-loop (Line 3–Line 10) of Algorithm n-SEPA can be executed in polynomial time.*

Using complete representative sets of cheapest alternatives, it is also possible to compute smallest tight alternatives in polynomial time:

Lemma 3.4.13 *Let $\mathcal{N} = (G, (s_i, t_i)_{i\in N}, c, d)$ be a network, where $(G, (s_i, t_i)_{i\in N})$ is n-series-parallel. Furthermore, let P be a strategy profile and $(\xi_{i,e})_{i\in N, e\in P_i}$ an*

optimal LP(P)-solution. If a smallest tight alternative of player j for edge $f \in P_j$
exists, we can compute such an alternative in polynomial time.

Proof We show that a smallest tight alternative of player j for f can be obtained as follows. For a given complete representative set $\mathcal{A}_j$ of cheapest alternatives according to P, let $\mathcal{A}'_j$ denote all alternatives $A_j \in \mathcal{A}_j$ with $f \in S_{A_j}$
and $\sum_{e \in S_{A_j}} \left(\xi_{j,e} + d_{j,e} \right) = \sum_{e \in A_j} \left(c_e(n_e(P) + 1) + d_{j,e} \right)$. Furthermore, let A'_j
be the unique alternative in $\mathcal{A}'_j$ fulfilling $S_{A'_j} \subsetneq S_{A_j}$ for all $A_j \in \mathcal{A}'_j$. Then,
$P'_j := P_j \setminus S_{A'_j} \cup A'_j$ is a smallest tight alternative of player j for f: Obviously, P'_j
is a tight alternative for f, since $P'_j \in \mathcal{P}_j$, $f \notin P'_j$ and

$$
\sum_{e \in P_j \setminus P'_j} \left(\xi_{j,e} + d_{j,e} \right) = \sum_{e \in S_{A'_j}} \left(\xi_{j,e} + d_{j,e} \right)
$$
$$
= \sum_{e \in A'_j} \left(c_e(n_e(P) + 1) + d_{j,e} \right) = \sum_{e \in P'_j \setminus P_j} \left(c_e(n_e(P) + 1) + d_{j,e} \right).
$$

Furthermore, $P_j \cup P'_j$ contains a unique cycle (namely $S_{A'_j} \cup A'_j$). Finally, the
existence of a tight alternative P''_j for f with $P_j \setminus P''_j \subsetneq P_j \setminus P'_j$ is not possible,
since such an alternative yields an alternative $A''_j := P''_j \setminus P_j \in \mathcal{A}'_j$ with $S_{A''_j} \subsetneq S_{A'_j}$,
contradicting the choice of A'_j. Thus, the described procedure yields a smallest tight
alternative of player j for f, and obviously, all steps can be executed in polynomial
time.   □

Using Lemma 3.4.13, we conclude that in Line 21 of Algorithm n-SEPA, a smallest
tight alternative of player i for $\bar{E}_i$ can be computed in polynomial time (since
Algorithm TIGHT can be executed in polynomial time). This yields the following
result about the running time of the second inner repeat-loop.

Proposition 3.4.14 *A single iteration of the second inner repeat-loop (Line 18–*
Line 29) of Algorithm n-SEPA can be executed in polynomial time.

In the proof of Proposition 3.4.10, we showed that the number of iterations of the
second inner repeat-loop is $O(|E|)$ (in a fixed iteration of the outer repeat-loop). In
the following, we analyze the number of iterations of the outer, and the first inner
repeat-loop, in the special case with *constant* edge cost functions.

Proposition 3.4.15   *Let* $\mathcal{N} = (G, (s_i, t_i)_{i \in N}, c, d)$ *be a network, where* $(G, (s_i, t_i)_{i \in N})$ *is n-series-parallel and all edge cost functions* $c = (c_e)_{e \in E}$ *are constant functions. Then in a fixed iteration of the outer repeat-loop of Algorithm n-SEPA, there are* $O(n \cdot (|V| + |E|))$ *iterations of the first inner repeat-loop (Line 3–Line 10).*

Proof   Let $\bar{P}$ be the assignment of the variable P in the beginning of a fixed iteration of the outer repeat-loop. Then, in an arbitrary iteration of the first inner repeat-loop in which LP(P) is not feasible,

$$\{v \in V : v \in \bar{P}_i \setminus P_i\} \cup \{e \in E : e \in \bar{P}_i \setminus P_i\} \subsetneq \{v \in V : v \in \bar{P}_i \setminus P_i'\} \cup \{e \in E : e \in \bar{P}_i \setminus P_i'\} \tag{3.6}$$

holds for the player i who deviates from P_i to P_i', that is, the set of nodes and edges of $\bar{P}_i$ which are not used anymore by player i strictly increases (we prove this below). Obviously, this shows the stated bound on the number of iterations of the first inner repeat-loop.

We now prove the above stated property. Assume, by contradiction, that there is an iteration (of the first inner repeat-loop) in which (3.6) does not hold. Consider the *first* such iteration (in the following denoted as the *current* iteration), and let i be the player who deviates in this iteration. Note that the current iteration cannot be the first iteration with respect to the property that player i deviates, thus $P_i \neq \bar{P}_i$ holds (if $P_i = \bar{P}_i$ holds, (3.6) is satisfied since at least one edge of P_i is substituted). We now introduce some notation needed in the proof (see Figure 3.84 for illustration). Denote the subpaths of $\bar{P}_i$ which are edge-disjoint and internal node-disjoint with P_i, and are unextendable with respect to this property, by $S_1, \ldots, S_j$. For $k = 1, \ldots, j$, let u_k and v_k be the endnodes of S_k, where u_k is before v_k if we consider $\bar{P}_i$ from s_i to t_i, and A_k denotes the subpath of P_i with endnodes u_k and v_k.

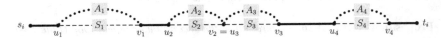

Fig. 3.84 Illustration for the proof of Proposition 3.4.15: $\bar{P}_i$ consists of the thick and the dashed subpaths, P_i of the thick and the dotted subpaths

Let u and v be the endnodes of the subpath $A := P_i' \setminus P_i$, where u is before v if we consider P_i from s_i to t_i. We now show that there is a $k \in \{1, \ldots, j\}$ such that u and v are both in $V(A_k)$, where $V(A_k)$ denotes the nodes which are used in

A_k. First of all, u and v are both contained in $\bigcup_{k=1}^{j} V(A_k)$: Otherwise, the series-parallel structure of $\mathcal{P}_i$ implies that at least one additional edge of $\bar{P}_i$ is substituted in the current iteration (see Figure 3.85a for an example), and thus (3.6) is fulfilled (contradiction). If there is no $k \in \{1, \ldots, j\}$ such that u and v are both in $V(A_k)$, the series-parallel structure together with the assumption that (3.6) does not hold yield the existence of k and k' with $k < k'$, such that $u = u_k$ and $v = v_{k'}$. But then (3.6) again holds, since at least one additional node or edge of $\bar{P}_i$ is substituted (see Figure 3.85b for the case that no edge, but an additional node, namely v_2, is substituted).

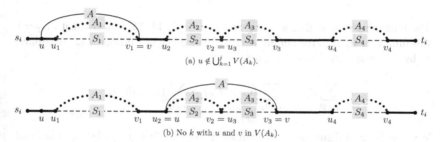

(a) $u \notin \bigcup_{k=1}^{j} V(A_k)$.

(b) No k with u and v in $V(A_k)$.

Fig. 3.85 Situations in the proof of Proposition 3.4.15, in which (3.6) holds

We can thus assume that u and v are both in $V(A_k)$, for a $k \in \{1, \ldots, j\}$. Due to our assumption that the current iteration is the *first* iteration in which (3.6) is violated, there exists an iteration (previous to the current one) in which player i substituted a path with endnodes u_k and v_k by A_k. Denote the substituted path with S'_k. In the following, we derive a contradiction using the fact that player i used a *minimally* violated path in this iteration. We distinguish between two cases.

In the first case, $u \neq u_k$ or $v \neq v_k$ holds. The series-parallel structure then implies that A is edge- and internal node-disjoint with S'_k. Thus if we denote the subpaths of A_k from u_k to u, from u to v, and from v to v_k with A_k^1, A_k^2 and A_k^3, then the path consisting of A_k^1, A and A_k^3 also substitutes S'_k (see Figure 3.86a for illustration). Since player i used a minimally violated path, the costs plus delays for this path are not smaller than the costs plus delays for A_k (cf. 3. in Definition 3.4.2), that is,

$$\sum_{e \in A_k^1 \cup A \cup A_k^3} (c_e + d_{i,e}) \geq \sum_{e \in A_k = A_k^1 \cup A_k^2 \cup A_k^3} (c_e + d_{i,e}) \Leftrightarrow \sum_{e \in A} (c_e + d_{i,e}) \geq \sum_{e \in A_k^2} (c_e + d_{i,e}),$$

where $c_e \geq 0$ denotes the constant value of the edge cost function c_e of $e \in E$. On the other hand, player i substitutes A_k^2 by A in the current iteration, thus

$$\sum_{e \in A_k^2} d_{i,e} > \sum_{e \in A} (c_e + d_{i,e})$$

holds. Putting the inequalities together yields the following contradiction:

$$\sum_{e \in A_k^2} d_{i,e} > \sum_{e \in A} (c_e + d_{i,e}) \geq \sum_{e \in A_k^2} (c_e + d_{i,e}) \geq \sum_{e \in A_k^2} d_{i,e}$$

The remaining case is that $u = u_k$ and $v = v_k$ hold. This means that player i substitutes A_k by A in the current iteration. Since player i furthermore substituted S_k' by A_k (in an earlier iteration), we get

$$\sum_{e \in S_k'} d_{i,e} > \sum_{e \in A_k} (c_e + d_{i,e}) > \sum_{e \in A} (c_e + d_{i,e}).$$

We now have to distinguish between the two cases that A is edge- and internal node-disjoint with S_k', or not. In the former case, we can argue as before: Since A is an alternative for S_k', and player i used a *minimally* violated path, one gets from 3. in Definition 3.4.2 that $\sum_{e \in A} (c_e + d_{i,e}) \geq \sum_{e \in A_k} (c_e + d_{i,e})$ holds, contradicting $\sum_{e \in A_k} (c_e + d_{i,e}) > \sum_{e \in A} (c_e + d_{i,e})$ from above. If, otherwise, A is *not* edge- and internal node-disjoint with S_k' (see Figure 3.86b for an illustrative example), we can derive

$$\sum_{e \in S_k' \setminus A} d_{i,e} \leq \sum_{e \in A \setminus S_k'} (c_e + d_{i,e})$$

as follows. In the iteration in which player i substituted S_k' by A_k (using a *minimally* violated path), there was no violated path substituting a strict subset of S_k' (cf. 2. in Definition 3.4.2; in Figure 3.86b, the delays of a dashed path are not larger than the costs plus delays of the dotted path having the same endnodes). This (together with the series-parallel structure) implies the stated inequality. Putting everything together yields

$$\sum_{e \in S'_k} d_{i,e} > \sum_{e \in A} (c_e + d_{i,e}) = \sum_{e \in A \cap S'_k} (c_e + d_{i,e}) + \sum_{e \in A \setminus S'_k} (c_e + d_{i,e})$$

$$\geq \sum_{e \in S'_k \cap A} d_{i,e} + \sum_{e \in S'_k \setminus A} d_{i,e} = \sum_{e \in S'_k} d_{i,e},$$

which is a contradiction and thus completes the proof. □

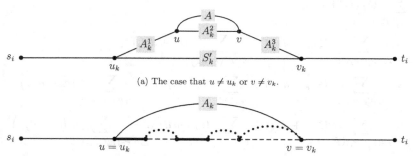

(a) The case that $u \neq u_k$ or $v \neq v_k$.

(b) The case that $u = u_k$, $v = v_k$, and A is not edge- and internal node-disjoint with S'_k (A is given by the thick and the dotted paths; and S'_k by the thick and the dashed paths).

Fig. 3.86 Remaining cases in the proof of Proposition 3.4.15

Proposition 3.4.16 *Let* $\mathcal{N} = (G, (s_i, t_i)_{i \in N}, c, d)$ *be a network, where* $(G, (s_i, t_i)_{i \in N})$ *is n-series-parallel and all edge cost functions* $c = (c_e)_{e \in E}$ *are constant functions. Then Algorithm n-*SEPA *terminates in the first iteration of the outer repeat-loop.*

Proof Consider the first iteration of the outer repeat-loop and let P and $(\xi_{i,e})_{i \in N, e \in P_i}$ be the strategy profile and the cost shares computed in the first inner repeat-loop (Line 3–Line 10). If P is enforceable, the variable enforceable is set to *true* in Line 14 and we terminate. Thus assume that P is not enforceable. Let P' be the strategy profile computed in the second inner repeat-loop (Line 18–Line 29), and let $(\xi_{i,e}(P'))_{i \in N, e \in P'_i}$ be the corresponding cost shares after the possible reduction in Line 30. It remains to show that $(\xi_{i,e}(P'))_{i \in N, e \in P'_i}$ is a feasible solution for LP(P') (then, enforceable is set to *true* in Line 32 and we terminate). Since all used edges are exactly paid and the cost shares are nonnegative, we only need to verify condition (2) of LP(P'), i.e., that

$$\sum_{e\in P'_j\backslash\bar{P}_j}\left(\xi_{j,e}(P')+d_{j,e}\right)\le\sum_{e\in\bar{P}_j\backslash P'_j}\left(c_e+d_{j,e}\right)$$

holds for each $j\in N$, $\bar{P}_j\in\mathcal{P}_j$, where $c_e\ge 0$ denotes the constant value of the edge cost function c_e of $e\in E$. The desired inequality can be shown as follows (explanations below):

$$
\begin{aligned}
&\sum_{e\in P'_j\backslash\bar{P}_j}\left(\xi_{j,e}(P')+d_{j,e}\right)\\
\le\;&\sum_{e\in(P'_j\cap P_j)\backslash\bar{P}_j}\left(\xi_{j,e}+d_{j,e}\right)+\sum_{e\in(P'_j\backslash P_j)\backslash\bar{P}_j}\left(c_e+d_{j,e}\right)\\
=\;&\sum_{e\in(P'_j\cap P_j)\backslash\bar{P}_j}\left(\xi_{j,e}+d_{j,e}\right)+\sum_{e\in P'_j\backslash P_j}\left(c_e+d_{j,e}\right)-\sum_{e\in(P'_j\cap\bar{P}_j)\backslash P_j}\left(c_e+d_{j,e}\right)\\
=\;&\sum_{e\in(P'_j\cap P_j)\backslash\bar{P}_j}\left(\xi_{j,e}+d_{j,e}\right)+\sum_{e\in P_j\backslash P'_j}\left(\xi_{j,e}+d_{j,e}\right)-\sum_{e\in(P'_j\cap\bar{P}_j)\backslash P_j}\left(c_e+d_{j,e}\right)\\
=\;&\sum_{e\in(P'_j\cap P_j)\backslash\bar{P}_j}\left(\xi_{j,e}+d_{j,e}\right)+\sum_{e\in P_j\backslash\bar{P}_j}\left(\xi_{j,e}+d_{j,e}\right)-\sum_{e\in(P_j\cap P'_j)\backslash\bar{P}_j}\left(\xi_{j,e}+d_{j,e}\right)\\
&+\sum_{e\in(P_j\cap\bar{P}_j)\backslash P'_j}\left(\xi_{j,e}+d_{j,e}\right)-\sum_{e\in(P'_j\cap\bar{P}_j)\backslash P_j}\left(c_e+d_{j,e}\right)\\
\le\;&\sum_{e\in\bar{P}_j\backslash P_j}\left(c_e+d_{j,e}\right)+\sum_{e\in(P_j\cap\bar{P}_j)\backslash P'_j}\left(\xi_{j,e}+d_{j,e}\right)-\sum_{e\in(P'_j\cap\bar{P}_j)\backslash P_j}\left(c_e+d_{j,e}\right)\\
=\;&\sum_{e\in\bar{P}_j\backslash P'_j}\left(c_e+d_{j,e}\right)+\sum_{e\in(P'_j\cap\bar{P}_j)\backslash P_j}\left(c_e+d_{j,e}\right)-\sum_{e\in(P_j\cap\bar{P}_j)\backslash P'_j}\left(c_e+d_{j,e}\right)\\
&+\sum_{e\in(P_j\cap\bar{P}_j)\backslash P'_j}\left(\xi_{j,e}+d_{j,e}\right)-\sum_{e\in(P'_j\cap\bar{P}_j)\backslash P_j}\left(c_e+d_{j,e}\right)\\
\le\;&\sum_{e\in\bar{P}_j\backslash P'_j}\left(c_e+d_{j,e}\right)
\end{aligned}
$$

The first inequality follows from the definition of $(\xi_{i,e}(P'))_{i\in N,e\in P'_i}$ together with the possible reduction of the cost shares in Line 30, the second equality from the tightness of P'_j, and the following inequalities from the feasibility of $(\xi_{i,e})_{i\in N, e\in P_i}$ for LP(P). All other equalities are due to simple set calculus. $\qquad\square$

If the edge cost functions are not constant, but load-dependent, it is not true in general that the cost shares $(\xi_{i,e}(P'))_{i\in N, e\in P'_i}$ computed in Line 30 of Algorithm *n*-SEPA fulfill condition (2) of LP(P'). In particular, we can not guarantee that Algorithm *n*-SEPA terminates in the first iteration of the outer repeat-loop. This is illustrated in the following example, which concludes the subsection.

Example 3.4.17 Consider the network displayed in Figure 3.87a. It is easy to verify that $(G, (s_i, t_i)_{i\in\{1,2\}})$ is 2-series-parallel and that all edge cost functions are nondecreasing and discrete-concave. All delays are assumed to be zero. We now apply Algorithm *n*-SEPA with the strategy profile $P = (P_1, P_2)$, defined by the thick edges in Figure 3.87b, as input profile. Figure 3.87b also shows an optimal LP(P)-solution. Note that the commonly used edge f is not completely paid ($\xi_{1,f} + \xi_{2,f} = 6 + 3 = 9 < 10 = c_f(2)$), thus P is not enforceable. Figure 3.87c shows the strategy profile $P' = (P'_1, P'_2)$ and the corresponding cost shares $(\xi_{i,e}(P'))_{i\in\{1,2\}, e\in P'_i}$ computed in the second inner repeat loop (Line 18–Line 29). Note that all used edges are exactly paid, thus Line 30 does not change the cost shares. But the cost shares are *not feasible* for LP(P'), since condition (2) of LP(P') is not fulfilled: For the path P_1 which Player 1 used in the input profile P, we get

$$\sum_{e\in P'_1 \setminus P_1} \xi_{1,e}(P') = 8 > 7 = 1 + 5 + 1 = \sum_{e\in P_1 \setminus P'_1} c_e((P_1, P'_2)).$$

Algorithm *n*-SEPA thus updates P with P' and starts a second iteration of the outer repeat-loop. For completeness, Figure 3.87d shows the strategy profile P'' which is computed in this iteration. Since the displayed cost shares are now feasible, this is also the output of Algorithm *n*-SEPA.

3.4.3 Main Results

In this subsection, we formulate and prove our main results for *n*-player network cost sharing games.

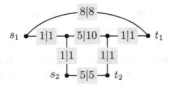

(a) Network $(G, (s_i, t_i)_{i \in \{1,2\}}, c, d)$ of the example: Edges $e \in E$ are labelled with $c_e(1)|c_e(2)$ and all delays are zero.

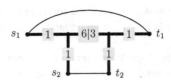

(b) Input profile P for Algorithm n-SEPA; used edges are labelled with an optimal LP(P)-solution.

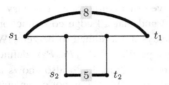

(c) Profile P' computed in the second inner repeat-loop; used edges are labelled with the corresponding cost shares.

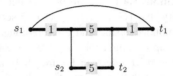

(d) Output profile P'' of Algorithm n-SEPA; used edges are labelled with the corresponding cost shares.

Fig. 3.87 Illustration for Example 3.4.17

Theorem 3.4.18 *Let $\mathcal{N} = (G, (s_i, t_i)_{i \in N}, c, d)$ be a network, where $(G, (s_i, t_i)_{i \in N})$ is n-series-parallel and all edge cost functions $c = (c_e)_{e \in E}$ are nondecreasing and discrete-concave. Then using an arbitrary strategy profile $P \in \mathcal{P}$ as input, Algorithm n-SEPA returns P if P is enforceable, and else it computes an enforceable strategy profile $P' \in \mathcal{P}$ with $C(P') < C(P)$.*

Proof Due to Subsection 3.4.2.2 and Proposition 3.4.10, we know that Algorithm n-SEPA is well-defined and terminates with strategy profile P'. Proposition 3.4.7 then shows the desired properties of the output profile P'.   □

The following Corollary 3.4.19 is a simple consequence of Theorem 3.4.18: If the input profile P is not enforceable, Algorithm n-SEPA computes a strategy profile with strictly smaller cost. Since this leads to a contradiction if P additionally is an *optimal* strategy profile, we conclude that every optimal strategy profile is enforceable.

Corollary 3.4.19 Let $\mathcal{N} = (G, (s_i, t_i)_{i \in N}, c, d)$ be a network, where $(G, (s_i, t_i)_{i \in N})$ is n-series-parallel and all edge cost functions $c = (c_e)_{e \in E}$ are non-decreasing and discrete-concave. Then every optimal strategy profile of the network $\mathcal{N}$ is enforceable. Or, in other words, an n-series-parallel graph $(G, (s_i, t_i)_{i \in N})$ is strongly efficient for nondecreasing, discrete-concave edge costs and arbitrary (fixed) delays.

If all edge cost functions are constant, we can guarantee polynomial running time for Algorithm n-SEPA.

Theorem 3.4.20 Let $\mathcal{N} = (G, (s_i, t_i)_{i \in N}, c, d)$ be a network, where $(G, (s_i, t_i)_{i \in N})$ is n-series-parallel and all edge cost functions $c = (c_e)_{e \in E}$ are constant functions. Then Algorithm n-SEPA can be executed in polynomial time.

Proof Recall that single iterations of the two inner repeat-loops can be executed in polynomial time (see Proposition 3.4.12 and Proposition 3.4.14). Furthermore, in a fixed iteration of the outer repeat-loop, the number of iterations of the first inner repeat-loop is $O(n \cdot (|V| + |E|))$ (see Proposition 3.4.15), and the number of iterations of the second inner repeat-loop is $O(|E|)$ (see the proof of Proposition 3.4.10). Altogether, we conclude that a single iteration of the outer repeat-loop can be executed in polynomial time (note that the feasibility of $(\xi_{i,e}(P'))_{i \in N, e \in P_i'}$ for LP(P') (in Line 31) can be tested in polynomial time by using the linear program LP$(P')'$ introduced in the proof of Lemma 3.4.11). Since we showed in Proposition 3.4.16 that Algorithm n-SEPA already terminates in the *first* iteration of the outer repeat-loop, the desired claim follows.   □

Using Theorem 3.4.18 and Theorem 3.4.20, we can reduce the efficient computation of a cost sharing method with low PoS to the efficient computation of a low-cost strategy profile (thus, to the design of a good approximation algorithm for the underlying optimization problem): Assume that we use a strategy profile P with cost $C(P) \leq \alpha \cdot \text{OPT}$, where $\alpha \geq 1$ and OPT denotes the cost of a cost-minimal strategy profile, as input for Algorithm n-SEPA. If all edge cost functions are constant, Algorithm n-SEPA returns *in polynomial time* a strategy profile P' with $C(P') \leq C(P)$, and a cost sharing method ξ (via Theorem 3.2.1) such that P' is a PNE for the game induced by ξ. Thus, the PoS is at most $\frac{C(P')}{\text{OPT}} \leq \frac{C(P)}{\text{OPT}} \leq \alpha$. This is the statement of the following corollary.

Corollary 3.4.21 Let $\mathcal{N} = (G, (s_i, t_i)_{i \in N}, c, d)$ be a network, where $(G, (s_i, t_i)_{i \in N})$ is n-series-parallel and all edge cost functions $c = (c_e)_{e \in E}$ are

constant *functions. If we can efficiently compute a strategy profile P with cost*
$C(P) \leq \alpha \cdot OPT$, *where* $\alpha \geq 1$ *and OPT denotes the cost of an optimal strategy
profile, then we can efficiently compute a cost sharing method* ξ *with* $PoS(\xi) \leq \alpha$.

A natural question arising from Corollary 3.4.21 is, if computing an optimal, or near-
optimal, strategy profile is always possible in polynomial time. In general, unless
$\mathcal{P} = \mathcal{NP}$, we cannot efficiently compute an optimal strategy profile. This follows
from the fact that the uncapacitated facility location problem (UFL) is contained as
a special case, and UFL is known to be $\mathcal{NP}$-hard to solve (see Subsection 3.4.4).
Furthermore, all inapproximability results of UFL also carry over (see also Subsection 3.4.4). However, for the special case of constant edge costs and *zero* delays in
an *undirected* n-series-parallel graph, it is possible to compute an optimal strategy
profile in polynomial time. We prove this in the next proposition, where we make use
of the result of Bateni et al. [12], that an optimal Steiner forest (cf. Subsection 3.3.1)
can be computed efficiently for a graph with *treewidth* at most two (see the proof
of Proposition 3.4.22 for a definition of the treewidth of a graph).

Proposition 3.4.22 *Let* $\mathcal{N} = (G, (s_i, t_i)_{i \in N}, c, d)$ *be a network, where*
$(G, (s_i, t_i)_{i \in N})$ *is an* undirected *n-series-parallel graph, all edge cost functions*
$c = (c_e)_{e \in E}$ *are* constant *functions and all delays* $d = (d_{i,e})_{i \in N, e \in E}$ *are zero. Then
an optimal strategy profile can be computed in polynomial time.*

Proof Note that if all edge cost functions are constant and all delays are zero, every
optimal Steiner forest yields an optimal strategy profile (cf. Subsection 3.3.1). Thus
it suffices to compute an optimal Steiner forest. In Bateni et al. [12], it is shown
that an optimal Steiner forest can be computed in polynomial time if the underlying
graph has *treewidth* at most two (below, we will give a definition for the treewidth
of a graph). Using this result, our proof strategy is as follows: First, we argue that
all nodes and edges with the property that there is no path in the union of all (s_i, t_i)-
paths $\cup_{i \in N} \mathcal{P}_i$ containing them, can be found and deleted from G in polynomial time
(it is clear that this does not change the set of (optimal) strategy profiles or Steiner
forests). Next, we show that each connected component of the resulting graph is
generalized series-parallel, that is: It can be created by a sequence of *series*, *parallel*
and *add* operations starting from a single (undirected) edge, where an add operation
adds a new vertex u and connects it to an existing vertex v by the edge $u - v$.
Finally, we show that a generalized series-parallel graph has treewidth at most two.
Altogether, this shows Proposition 3.4.22.

Finding and deleting all nodes and edges which are not used by the paths in
$\cup_{i \in N} \mathcal{P}_i$ is possible via a result of Chen et al. [26]. There, a subgraph H of G is

called *irredundant* if each node and each edge of H is contained in at least one (s_i, t_i)-path in H for some $i \in N$, and the *maximum* irredundant subgraph is the irredundant subgraph having the largest number of edges. Using these notions, our goal is to compute the maximum irredundant subgraph of G. It is quite clear (and shown in [26]), that the maximum irredundant subgraph can be computed in polynomial time as follows (cf. Algorithm 1 in [26]). For $i \in N$, set $G_i := G$ and delete from G_i all connected components except the one containing s_i and t_i. Let C_i be the set of cut vertices of G_i (recall that a *cut vertex* is a vertex c having the property that if we delete c from G_i, the resulting subgraph, denoted by $G_i - c$, has a larger number of connected components than G_i). For each cut vertex $c \in C_i$, consider the connected components of $G_i - c$. If there are components containing neither s_i nor t_i, delete from G_i all vertices which are contained in these components. The union of all G_is computed this way is the maximum irredundant subgraph of G.[15]

We now show that each connected component of the maximum irredundant subgraph of G is generalized series-parallel. In fact, it suffices to show the statement for the case that G is irredundant and connected (note that each connected component of the maximum irredundant subgraph of G obviously induces an n'-series-parallel graph, for some subset N' of the players with $|N'| = n'$). Thus we now assume that G is irredundant and connected, and we show that G can be created by a sequence of series, parallel and add operations starting from a single edge. Since $(G, (s_i, t_i)_{i \in N})$ is n-series-parallel, it is clear that for each $i \in N$, the subgraph G_i induced by $\mathcal{P}_i$ can be created like this. But it is not clear a priori whether this also holds for G, the union of all G_is (note that, e.g., there may be G_i and G_j such that neither is a subgraph of the other, but they do contain common nodes or edges). We now show, starting with the subgraph G_1 which is generalized series-parallel, that we can consecutively choose one player and add the nodes and edges of her paths which are not already contained in the subgraph constructed so far by series, parallel and add operations. Since this again yields a generalized series-parallel graph, we finally conclude that G is generalized series-parallel. Let $G' \neq G$ be the generalized series-parallel subgraph constructed so far, and $N' \subsetneq N$ the set of players whose paths are already added to G'. Choose a player $i \in N \setminus N'$ such that G_i is not node-disjoint with G' (such a player exists since G is connected). Let P_i be an (s_i, t_i)-path which is not node-disjoint with G' and subdivide P_i into the three subpaths P_i^1, P_i^2, P_i^3

---

[15]Note that for *directed* graphs, there is no efficient algorithm computing the maximum irredundant subgraph (unless $\mathcal{P} = \mathcal{NP}$): This follows from the fact that deciding whether a given directed edge $u \rightarrow v$ is contained in an (s_i, t_i)-path, is equivalent to deciding whether there are vertex-disjoint (s_i, u)- and (v, t_i)-paths, and the latter problem is known to be $\mathcal{NP}$-complete [46].

(where some of the subpaths may consist of only one node) as follows: Considering P_i from s_i to t_i, let u (v) be the first (last) node of P_i which is contained in G'. Then, P_i^1 is the subpath of P_i from s_i to u, P_i^2 is the subpath from u to v, and P_i^3 is the subpath from v to t_i. Note that G_i consists of P_i together with all alternatives of player i according to P_i (cf. page 157, where we introduced the notion of alternatives according to P_i). In the following, we show that all nodes and edges of G_i which are not already contained in G' can be added to G' by series, parallel and add operations. We distinguish between $u \neq v$ and $u = v$, and use the series-parallel structure of G_i together with the fact that there cannot be a new (s_j, t_j)-path for a player $j \in N'$. First, consider the case that $u \neq v$, i.e., P_i^2 contains at least one edge. The subpath P_i^2, as well as all alternatives with both endnodes in P_i^2, are already contained in G'. Furthermore, any alternative which does not have both endnodes in P_i^2 either has both endnodes in P_i^1, or both endnodes in P_i^3, and is internal node-disjoint with G'. Therefore these alternatives, as well as P_i^1 and P_i^3, can be added to G' with at most two add operations at u and v, followed by series and parallel operations (see Figure 3.88a for an illustrative example). Now turn to the case that $u = v$. In this case, any alternative is internal node-disjoint with G', and has at least one endnode which is different from u (that is, which is not contained in P_i^2). If each alternative either has both endnodes in P_i^1, or both endnodes in P_i^3, we can obviously add P_i^1, P_i^3 and all alternatives to G' by at most two add operations at u, followed by series and parallel operations (as above). But there may also be alternatives which neither have both endnodes in P_i^1, nor both endnodes in P_i^3. That is, one endnode is in P_i^1, one endnode is in P_i^3, and both endnodes are different from u. In particular, such an alternative substitutes u. In this case, we first add all alternatives substituting u, as well as the edges of P_i which are substituted by them, by an add operation at u, followed by series and parallel operations. After that, we can add the remaining alternatives and edges of P_i by at most two add operations, followed by series and parallel operations (as before). Figure 3.88b shows an illustrative example for this situation. Altogether, we thus showed that G is generalized series-parallel.

To complete the proof, it remains to show that a generalized series-parallel graph has treewidth at most two. Using induction on the number of series, parallel and add operations needed to generate the graph, this is a straightforward exercise, but for completeness, we now present a proof. First, we define the treewidth of a graph as follows (see [12]). A *tree decomposition* of a graph $G = (V, E)$ is a pair $(T, \mathcal{B})$ in which $T = (I, F)$ is a tree and $\mathcal{B} = \{B_i : i \in I\}$ is a family of subsets of V such that the following three properties hold: (1) $\bigcup_{i \in I} B_i = V$; (2) For each edge $u - v \in E$, there exists an $i \in I$ such that both u and v belong to B_i; and (3) For each $v \in V$, the subgraph of T induced by $\{i \in I : v \in B_i\}$ is a tree. The *width* of a tree decomposition $(T, \mathcal{B})$ is the maximum size of an element $B \in \mathcal{B}$ minus 1, i.e.,

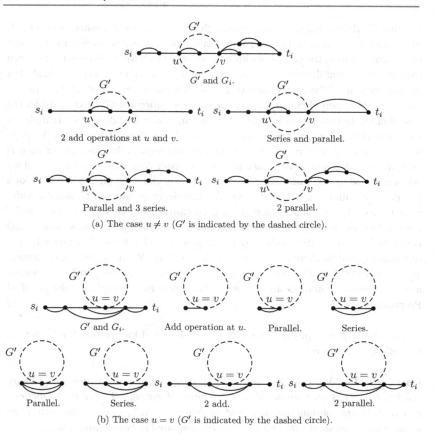

(a) The case $u \neq v$ (G' is indicated by the dashed circle).

(b) The case $u = v$ (G' is indicated by the dashed circle).

Fig. 3.88 Illustration for the proof of Proposition 3.4.22

$\max\{|B| : B \in \mathcal{B}\} - 1$. The *treewidth* of a graph G is the minimum width over all possible tree decompositions of G. Now assume, starting with the single edge $s - t$, that there is a sequence of series, parallel and add operations creating G, and the total number of operations in this sequence is $k \geq 0$. We prove that G has treewidth at most two by induction on k. The base case is $k = 0$, that is, G consists of the single edge $s - t$ only. Obviously, the tree consisting of the single node 1, together with $\mathcal{B} = \{B_1 = \{s, t\}\}$, constitute a tree decomposition with width 1, showing the claim for the base case. For the induction step, let G' be the graph resulting from the first $k - 1$ operations in the sequence of series, parallel and add operations creating G. By the induction hypothesis, there exists a tree decomposition $(T, \mathcal{B})$ of G' with width at most two. We now adapt this tree decomposition to achieve a tree decomposition for G with width at most two. We have to distinguish between the three cases that the last operation (creating G from G') is a series, a parallel or an add operation. The easiest case is that the last operation is a parallel operation, since $(T, \mathcal{B})$ is then obviously also a tree decomposition for G. Next, consider the case that the last operation is a series operation, where the edge $u - v$ is replaced by a new node w and two new edges $u - w$, $w - v$. Since G' contains the edge $u - v$, the second property of a tree decomposition implies the existence of $B_i \in \mathcal{B}$ containing both u and v. We now add a new element $B_j := \{u, v, w\}$ to $\mathcal{B}$, and we add a new node j and a new edge $i - j$ to T. By checking the three conditions for a tree decomposition, we get that this yields a tree decomposition of G, and the width is two since $|B_j| = 3$ and no element of $\mathcal{B}$ contains more than three nodes. Finally, assume that the last operation is an add operation which adds a new node u and connects it to an existing node v by the edge $u - v$. Due to the first property of a tree decomposition, there exists a $B_i \in \mathcal{B}$ with $v \in B_i$. We now add a new element $B_j := \{u, v\}$ to $\mathcal{B}$, and we add a new node j and a new edge $i - j$ to T. This yields a tree decomposition of G and the width is at most two, completing the proof of Proposition 3.4.22. $\qquad\square$

As a consequence of the above Corollary 3.4.21 and Proposition 3.4.22, we get the following corollary.

Corollary 3.4.23 *Let $\mathcal{N} = (G, (s_i, t_i)_{i \in N}, c, d)$ be a network, where $(G, (s_i, t_i)_{i \in N})$ is an* undirected *n-series-parallel graph, all edge cost functions $c = (c_e)_{e \in E}$ are constant functions and all delays $d = (d_{i,e})_{i \in N, e \in E}$ are zero. Then a cost sharing method with PoS of 1 can be computed in polynomial time.*

To conclude this subsection, we want to mention what one can achieve for *directed* n-series-parallel graphs, constant edge cost functions and zero delays. Assume that we have a network with a directed graph, constant edge cost functions and zero delays. To solve the underlying optimization problem, there are approximation algorithms achieving sublinear approximation guarantees in $|V|$, for example a guarantee of $O(|V|^{4/5+\varepsilon})$ due to [43], or an *expected* guarantee of $O(|V|^{2/3+\varepsilon})$ due to [16] (both for any constant $\varepsilon > 0$). Thus, by using one of these algorithms, we can efficiently compute cost sharing methods with PoS at most the stated guarantees (in networks consisting of a directed n-series-parallel graph, constant edge cost functions and zero delays).

3.4.4   Discussion of the Results

In this subsection, we briefly discuss the special setting of n-series-parallel graphs
for which we achieved our results for n-player network cost sharing games. We first
argue that although the assumption of an n-series-parallel graph is restrictive, network cost sharing games on n-series-parallel graphs contain nontrivial special cases,
e.g., *uncapacitated facility location (UFL)* cost sharing games. We furthermore show
that our results cannot be extended to the class of generalized series-parallel graphs.

   We start with the definition of UFL cost sharing games and show that these games
are special cases of network cost sharing games on n-series-parallel graphs.

Definition 3.4.24   (UFL problem/UFL cost sharing game). In the *UFL problem*,
we are given two finite, nonempty sets F and C, where the elements of F are called
facilities, and the elements of C are called customers. For each facility $f \in F$,
there is an opening cost $o_f \in \mathbb{R}_{\geq 0}$, and a distance $d_{c,f} \in \mathbb{R}_{\geq 0}$ to each customer
$c \in C$. A solution of the UFL problem is an assignment S which assigns, for each
customer $c \in C$, a facility $S(c) \in F$. The facilities which are assigned to at least one
customer need to be opened, and the customers experience delays measured by the
distance to their assigned facility. Thus, the overall cost of a solution S is defined as
$C(S) := \sum_{f \in F : S^{-1}(f) \neq \emptyset} o_f + \sum_{c \in C} d_{c,S(c)}$ (here, S^{-1} denotes the inverse image
of S). An optimal solution of the UFL problem is a cost-minimal solution.

   An *UFL cost sharing game* is a game-theoretic variant of the UFL problem in
which the customers are the players, and each player needs to choose exactly one
facility. That is, the strategy set of each player is given by the set F of facilities.
The strategy profiles are then exactly the solutions S of the UFL problem. If a
facility f is chosen by at least one player, the opening cost o_f is shared among
the players which chose this facility: For each $f \in F$, there is a cost sharing
method ξ_f which assigns, for each strategy profile S such that facility f is chosen
by at least one player, nonnegative cost shares $\xi_{c,f}(S) \in \mathbb{R}_{\geq 0}$ for each customer
c who chose facility f in S. The cost of player $c \in C$ under S is then defined as
$\xi_{c,S(c)}(S) + d_{c,S(c)}$, and each player seeks to minimize her own cost. Finally, the cost
of a strategy profile S is defined analogously to the cost of S in the UFL problem,
that is, $C(S) = \sum_{f \in F : S^{-1}(f) \neq \emptyset} o_f + \sum_{c \in C} d_{c,S(c)}$.

We now argue that UFL cost sharing games are special cases of network cost sha-
ring games on n-series-parallel graphs, where $n := |C|$. We describe two different
approaches: The first one uses an undirected graph and player-specific delays, the
second one a directed graph and zero delays. We start with the approach using an
undirected graph. For a given UFL cost sharing game, define the undirected graph

$G = (V, E)$ consisting of two vertices s and t and $|F|$ parallel edges $s - t$ (one edge $e_f = s - t$ for each facility $f \in F$). For each player $c \in C$, let $s_c := s$ and $t_c := t$. Obviously, $(G, (s_c, t_c)_{c \in C})$ is n-series-parallel, since $G_c = G$ for each player $c \in C$ and G can be constructed from the single edge $s - t$ by $|F| - 1$ parallel operations. For each edge $e_f \in E$, define the shareable edge cost as the opening cost o_f of facility f, and for each player $c \in C$, define the player-specific delay d_{c,e_f} as the distance $d_{c,f}$ of customer c to facility f. Finally, the cost sharing method of edge $e_f \in E$ is given by the cost sharing method ξ_f for facility $f \in F$. Obviously, the thus defined network cost sharing game is equivalent to the given UFL cost sharing game. Figure 3.89a shows the constructed undirected n-series-parallel graph with constant edge costs and player-specific delays. Figure 3.89b shows a different approach to model UFL cost sharing games as network cost sharing games: Here, we use a *directed* n-series-parallel graph defined as follows: The vertex set consists of a vertex s_c for each player $c \in C$, a vertex f for each facility $f \in F$, and a vertex t; the edge set contains, for each facility $f \in F$, a directed edge $f \to t$, and directed edges $s_c \to f$ for each player $c \in C$. The source-sink pair of player $c \in C$ is given by (s_c, t), and obviously, the graph G_c induced by the (s_c, t)-paths can be created from $s_c \to t$ by $|F| - 1$ parallel, followed by $|F|$ series operations. Thus, $(G, (s_c, t)_{c \in C})$ is n-series-parallel. Different to the previous approach with the undirected graph, we do not need player-specific delays here, that is, all delays are defined as zero. The constant shareable cost of edge $s_c \to f$ is defined as the distance $d_{c,f}$ of player $c \in C$ to facility $f \in F$, and for edge $f \to t$, the constant shareable cost is the opening cost o_f of facility $f \in F$. Finally, for each facility $f \in F$, the cost sharing method of edge $f \to t$ is given by the cost sharing method ξ_f, whereas the cost sharing method of $s_c \to f$ assigns the complete edge cost $d_{c,f}$ to player $c \in C$ if she uses this edge. As for the first approach, it is clear that the defined network cost sharing game is equivalent to the given UFL cost sharing game.

We showed that UFL cost sharing games are special cases of network cost sharing games on undirected (directed) n-series-parallel graphs with constant edge costs and with (without) player-specific delays. As a consequence, computing an optimal strategy profile is $\mathcal{NP}$-hard for these network cost sharing games: This follows from the fact that computing an optimal strategy profile of an UFL cost sharing game is equivalent to the UFL problem, and this problem is known to be $\mathcal{NP}$-hard to solve (since, for example, the set cover problem is a special case, see [80]).[16] Furthermore, all inapproximability results for the UFL problem also carry over

---

[16] In particular, this argument also shows that UFL cost sharing games cannot be modelled as network cost sharing games on *undirected* n-series-parallel graphs with constant edge costs

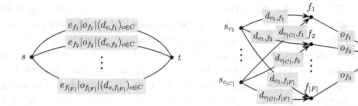

(a) Undirected n-series-parallel graph with constant edge costs and player-specific delays. Edges are labelled by their name, followed by their constant edge cost and their player-specific delays.

(b) Directed n-series-parallel graph with constant edge costs and zero delays. Edges are labelled with their constant edge cost.

Fig. 3.89 Modelling UFL as network cost sharing game

to the approximation of an optimal strategy profile: It is unlikely that there exists an approximation algorithm for the UFL problem with an approximation factor less than 1.463 for metric (see [57]), and $\ln(|C|)$ for general distances (derived for the set cover problem, see [42]). On the positive side, the (randomized) 1.488-approximation algorithm in [83] for the metric UFL problem, or the $O(\ln(|C|))$ greedy algorithm (see [68]) for the general case, lead by Corollary 3.4.21 to the efficient computation of cost sharing methods for UFL cost sharing games with PoS at most 1.488 for metric, or $O(\ln(|C|))$ for general (non-metric) distances.

We now show that our results can in general not be extended to the class of generalized series-parallel graphs. Recall that a generalized series-parallel graph is created by a sequence of series, parallel and add operations starting from a single undirected edge, where an add operation adds a new vertex u and connects it to an existing vertex v by the edge $u - v$. In the proof of Proposition 3.4.22, we showed that if the n-series-parallel graph $(G, (s_i, t_i)_{i \in N})$ additionally is undirected, connected, and has the property that any node and any edge of G is used by at least one player in one of her paths, then G is generalized series-parallel. This gives rise to the question whether our results can be extended to the class of generalized series-parallel graphs, in particular, whether this class also only contains strongly efficient graphs (for nondecreasing and discrete-concave edge costs and arbitrary delays). The following example shows that for $n \geq 3$ players, this does not hold, not even for constant edge costs and zero delays (whereas for $n = 2$, we already mentioned in Subsection 3.3.6 that every generalized series-parallel graph is strongly efficient for constant edge costs and zero delays).

---

and *zero* delays, since for these games, we can efficiently compute an optimal strategy profile (Proposition 3.4.22).

Example 3.4.25 Consider the network $\mathcal{N} = (G, (s_i, t_i)_{i \in N}, c, d)$ displayed in Figure 3.90a, where we have $n = 3$ players, the constant edge costs c are given on the edges, and all delays are zero. The unique optimal strategy profile OPT of $\mathcal{N}$ is induced by the thick edges and has cost $C(\text{OPT}) = 7 + 4 + 7 + 4 + 3 + 5 = 30$, but it is not enforceable: To see this, we use the LP-characterization from Section 3.2. We show that in any feasible solution for LP(OPT), the objective function value is strictly smaller than the cost of OPT, which shows the claim (by Theorem 3.2.1 and Observation 3.2.2). For all players, we now give upper bounds for the sum of cost shares that this player pays for the edges in her path in OPT. Player 1 pays at most 7, because she can use the edge $s_1 - t_1$ with cost 7. Player 3 can use the edge $s_3 - t_3$ with cost 8, thus she will pay at most 8. It remains to analyze the cost shares of Player 2. Instead of using the subpath from s_2 to s_1 of her path in OPT, Player 2 can use the edge $s_2 - s_1$ with cost 6. Furthermore, she can use the edge $t_3 - t_2$ with cost 8 instead of her subpath from t_3 to t_2. Since the two mentioned subpaths cover the complete path of Player 2 in OPT, the sum of Player 2's cost shares is at most $6 + 8 = 14$. Altogether, the objective function value of any feasible solution of LP(OPT) is at most $7 + 8 + 14 = 29 < 30 = C(\text{OPT})$, showing that OPT is not enforceable. Thus $(G, (s_i, t_i)_{i \in N})$ is not efficient for constant edge costs and zero delays. But G is generalized series-parallel, since it can be created by a sequence of series, parallel and add operations starting from a single edge (see Figure 3.90b).

To obtain a non-efficient, but generalized series-parallel graph for $n \geq 4$ players, we can for example add a new node u and connect it to s_1 by a new edge. Thus we executed an add operation and the resulting graph is generalized series-parallel. Setting $s_i := s_1$ and $t_i := u$ for all $i \in \{4, \dots, n\}$ obviously yields a graph which is not efficient for constant edge costs and zero delays (choosing an arbitrary nonnegative value for the constant cost of the new edge $u - s_1$ yields a network such that no optimal strategy profile is enforceable).

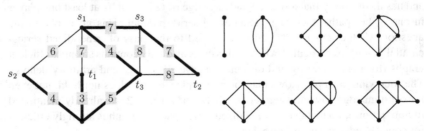

(a) Network $\mathcal{N}$ and optimal strategy profile (thick edges) used in the example.

(b) G is generalized series-parallel.

Fig. 3.90 Illustration for Example 3.4.25

Concluding, we mention a consequence of the above example. In Example 3.4.25, the unique optimal strategy profile has cost 30 and is not enforceable, and all edge costs are integral. Therefore, any enforceable strategy profile has cost at least 31, and a best separable and budget-balanced cost sharing method has PoS at least $31/30 \approx 1.033$. By optimizing the edge costs for the given graph, we can increase this lower bound to 1.043. This shows that the *worst-case* PoS of a best separable and budget-balanced cost sharing method, for a network cost sharing game played in a generalized series-parallel graph with constant edge costs and zero delays, is at least 1.043.

3.5 Bibliographic Notes

The results presented in Section 3.2 and Section 3.3 are part of a joint work with Tobias Harks and Manuel Surek, and are published in *SIAM Journal on Discrete Mathematics* [65]. Furthermore, an extended abstract appeared in the proceedings of the *13th International Conference on Web and Internet Economics (WINE 2017)* [62]. We now briefly describe to what extent this thesis complements the results published in [62, 65]. Since we consider more general cost sharing games in this thesis, the LP-characterization of enforceability presented in Section 3.2 is more general than the version in [62, 65] (there, we only considered enforceable Steiner forests in undirected graphs with constant edge costs and zero delays). Furthermore, the definition of efficiency and strong efficiency (for constant edge costs and zero delays) which is used in this thesis is slightly different, since in [62, 65], we used optimal and enforceable *Steiner forests* instead of strategy profiles. Note that by definition, using strategy profiles weakens efficiency, and strengthens strong efficiency. But as our characterization of efficiency (Theorem 3.3.1) shows, the definitions are in fact equivalent for the setting with two players, constant edge costs and zero delays in undirected graphs. To deal with the changed definition of efficiency, we added Subsection 3.3.1 in which we analyze the connection between strategy profiles and Steiner forests. Regarding the proof of our characterization of efficiency (Theorem 3.3.1), we significantly rewrote large parts of the proof and added further details and figures in order to make the proof more readable. In addition, some proofs and arguments are simplified, and proofs which are left to the reader in [65], are contained. Finally, we added a discussion of possible generalizations of our result in Subsection 3.3.6.

The results from Section 3.4 are part of a joint work with Tobias Harks, Martin Hoefer and Manuel Surek. An extended abstract of this work appeared in the proceedings of the *45th International Colloquium on Automata, Languages, and*

Programming (ICALP 2018) [60], and a full version is accepted for publication in *Mathematics of Operations Research* [61].[17] In the work mentioned above, we analyzed network cost sharing games in *undirected* n-series-parallel graphs with *constant* edge costs and delays. It is furthermore stated that the results can (partially) be generalized to include directed n-series-parallel graphs and nondecreasing, discrete-concave edge costs. In this thesis, we formulate and prove the results for this more general setting. Furthermore, to make the proofs more readable, we added further figures and details, and rearranged the structure of the technical parts of the paper.

Capacity and Price Competition in Networks 4

4.1 Introduction

In the last chapter, we analyzed a two-stage problem in which in the first stage, payments for using edges of a network are set, and in the second stage, selfish players use the edges and pay for it. In this chapter, we consider a problem motivated by privatized public roads, which also exhibits the described structure. In contrast to the problem of the last chapter, we now allow for a more complex setting in the first stage, but also impose more restrictions on the network structure. In particular, we assume that the prices are determined by a set of selfishly acting firms (and not, as before, by a central authority).

Consider a network which consists of a set of parallel edges connecting a source with a sink. Each edge corresponds to a firm owning the edge, and the firm sets a price for using her edge (but the price must not be above a given cap). Additionally, a firm can decide to invest in capacities at her edge, making it more "attractive" in the second stage selfish routing problem described as follows. After all firms have chosen capacities and prices for their edges, one unit of flow is to be routed from the source to the sink. Each flow particle represents an infinitesimally small customer wanting to travel from the source to the sink at minimum effective cost (price plus congestion cost). The congestion cost of an edge depends on the installed capacity and the volume of customers on that edge. Naturally, the congestion cost is decreasing in the capacity, and increasing in the volume of customers, and we particularly assume that the congestion cost is linear in the volume of customers, and inverse-linear in the capacity. Since each customer wants to minimize her own cost, the outcome of the described selfish routing problem can be assumed to be a Wardrop flow, a situation described by Wardrop's first principle [104], in which all flow travels along shortest paths. The profits of the firms can then be derived from

A. Schedel, *Cost Sharing, Capacity Investment and Pricing in Networks*, Mathematische Optimierung und Wirtschaftsmathematik | Mathematical Optimization and Economathematics, https://doi.org/10.1007/978-3-658-33170-2_4

the revenues gained in the underlying selfish routing problem, minus the capacity investment costs. The described game between the firms, which captures many aspects of realistic private road competition [107], is called a *capacity and price competition game*, and we analyze existence, uniqueness and quality of its PNE.

The organization of the chapter is as follows. In the remaining part of Section 4.1, we formally describe the problem, followed by a summary of our main results and proof techniques, and close with a brief discussion of the model and related work. The following sections then contain the technical presentation of our results. After analyzing the continuity of the profit functions (Section 4.2), we completely characterize the best response correspondences of the firms in Section 4.3. These two results are then used to derive results about existence (Section 4.4), uniqueness (Section 4.5) and quality (Section 4.6) of PNE. Finally, Section 4.7 contains some bibliographic notes.

4.1.1   Problem Description

In a *capacity and price competition game*, there is a set $N = \{1, \ldots, n\}$, $n \geq 2$, of firms (the players of the game). Each firm corresponds to an edge in a network, which consists of n parallel edges connecting a source with a sink, and a set of customers wants to travel from the source to the sink. Customers are represented by the continuum $[0, 1]$ (each consumer is assumed to be infinitesimally small and represented by a number in $[0, 1]$), and we denote by $P = \{x \in \mathbb{R}^n_{\geq 0} | \sum_{i \in N} x_i = 1\}$ the standard simplex of assignments of customers to firms.[1] The effective quality of firm i's edge—from the perspective of a customer—depends on two key factors: the level of congestion $\ell_i(x_i, z_i)$ and the price $p_i \geq 0$ charged by the firm. The congestion function $\ell_i(x_i, z_i)$ depends on the volume of customers x_i and the edge capacity $z_i \geq 0$ installed. Clearly $\ell_i(x_i, z_i)$ grows with the volume of customers x_i to be routed, but decreases with the edge capacity z_i. If no capacity is installed, i.e., $z_i = 0$, we assume infinite congestion, and for the case that $z_i > 0$, we assume that congestion depends *linearly* on the volume of customers and *inverse-linearly* on the installed capacity, that is,

$$\ell_i(x_i, z_i) = \begin{cases} \frac{a_i x_i}{z_i} + b_i, & \text{for } z_i > 0, \\ \infty, & \text{for } z_i = 0, \end{cases}$$

---

[1] All results hold for arbitrary intervals $[0, d]$, $d \in \mathbb{R}_{>0}$ by a standard scaling argument.

where $a_i > 0$ and $b_i \geq 0$ are given parameters for $i \in N$. Note that the case $z_i = 0$ can be interpreted as if firm i is just opting out of the game and her edge is not present in the network at all. Each firm i additionally decides on a price $p_i \in [0, C_i]$ which is charged to its customers for using the edge, where $C_i > 0$ is a given price cap. For a capacity vector $z = (z_1, \ldots, z_n)$ with $\sum_{i \in N} z_i > 0$, i.e., there is at least one edge present in the network, and a price vector $p = (p_1, \ldots, p_n)$, customers choose rationally the most attractive edge in terms of the *effective costs*, that is, congestion and price experienced. This is expressed by the *Wardrop equilibrium* conditions:

$$c_i(x, z, p) := \ell_i(x_i, z_i) + p_i \leq c_j(x, z, p) := \ell_j(x_j, z_j) + p_j$$

holds for all $i, j \in N$ with $x_i > 0$. Note that for given capacities $z \neq 0$ and prices p, there is exactly one $x \in P$ satisfying the Wardrop equilibrium conditions (see, e.g., [33]). Call this flow $x = x(z, p)$ the Wardrop flow *induced by* (z, p). In particular, there exists a constant $K \geq 0$ such that $c_i(x, z, p) = K$ holds for each $i \in N$ with $x_i > 0$, and $c_i(x, z, p) \geq K$ holds for each $i \in N$ with $x_i = 0$. For a Wardrop flow x, we call the corresponding constant K the *(routing) cost of x*. The profit function of a firm $i \in N$ can now be represented as

$$\Pi_i(z, p) = \begin{cases} p_i x_i(z, p) - \gamma_i z_i, & \text{for } \sum_{i \in N} z_i > 0, \\ 0, & \text{else,} \end{cases}$$

where $\gamma_i > 0$ is a given installation cost parameter for firm $i \in N$. We assume that each firm $i \in N$ seeks to maximize her own profit Π_i, and to this end chooses a strategy from her strategy set $S_i := \{s_i = (z_i, p_i) : 0 \leq z_i, 0 \leq p_i \leq C_i\}$. This setting is called a *capacity and price competition game*, and the objective of this chapter is to analyze existence, uniqueness and quality of these games' PNE.

At this point, we would like to draw the attention to a crucial assumption in the described setting. Namely, that the demand for routing is *inelastic*: As long as there exists a firm installing positive capacity, the volume of customers wanting to travel through the network stays constant. In particular, customers do not renounce travelling with increasing effective costs (for a discussion of this assumption, see Subsection 4.1.3). As already mentioned in Section 1.2, this assumption yields that standard approaches for showing existence of PNE can not be directly applied.

Concluding this subsection, we introduce some notations that we use throughout the chapter. As usual, a vector s consisting of strategies $s_i = (z_i, p_i) \in S_i$ for all $i \in N$ is called a strategy profile, and $S := \times_{i \in N} S_i$ denotes the set of strategy profiles. Usually, we will write a strategy profile $s \in \mathcal{S}$ in the form $s = (z, p)$, where z denotes the vector consisting of all capacities z_i for $i \in N$, and p is the vector of prices p_i for $i \in N$. The profit of firm i for a strategy profile $s = (z, p)$ is then defined as $\Pi_i(s) := \Pi_i(z, p)$. Furthermore, we write $x(s) := x(z, p)$ for the Wardrop-flow induced by $s = (z, p)$ and $K(s) := K(z, p)$ for the routing cost of $x(s)$. For firm i, denote by $s_{-i} = (z_{-i}, p_{-i}) \in \mathcal{S}_{-i} := \times_{j \in N \setminus \{i\}} \mathcal{S}_j$ the vector consisting of strategies $s_j = (z_j, p_j) \in \mathcal{S}_j$ for all $j \in N \setminus \{i\}$. We then write $(s_i, s_{-i}) = ((z_i, p_i), (z_{-i}, p_{-i}))$ for the strategy profile where firm i chooses $s_i = (z_i, p_i) \in \mathcal{S}_i$, and the other firms choose $s_{-i} = (z_{-i}, p_{-i}) \in \mathcal{S}_{-i}$. Moreover, we use the simplified notation $\Pi_i((s_i, s_{-i})) = \Pi_i(s_i, s_{-i})$ and $x((s_i, s_{-i})) = x(s_i, s_{-i})$.

4.1.2   Main Results and Proof Techniques

As our first result, we analyze under which conditions, for a given strategy profile, the profit functions of all firms are continuous at this strategy profile (namely if and only if at least one player installs some capacity, or all capacities and prices are zero, see Theorem 4.2.1). Second, we completely characterize the structure of best response correspondences of firms; including the possibility of non-existence of best responses (Theorem 4.3.6). We derive this result by showing that a best response of a firm exists if and only if at least one other firm has positive capacity, and in case of existence, the corresponding bilevel optimization problem can be solved by two ordinary, 1-dimensional optimization problems. Our first main result then establishes the existence of PNE (Theorem 4.4.6). Since the best response correspondence of a firm may have empty values, it is not clear how to apply Kakutani's fixed point theorem to show existence of PNE. Instead, we use the concept of C-security introduced by McLennan, Monteiro and Tourky [89], which in turn resembles ideas of Reny [95]. A game is C-secure at a given strategy profile, if each player has a strategy guaranteeing a certain utility value, even if the other players play some perturbed strategy within a (small enough) neighbourhood of the given strategy profile, and furthermore, for each slightly perturbed strategy profile, there is a player whose perturbed strategy can in some sense be strictly separated from her securing strategies. Intuitively, the concept of securing strategies means that those strategies are robust to other players' small deviations. The result of McLennan et al. states that a game with compact, convex strategy sets and bounded profit functions admits a PNE, if every non-equilibrium strategy profile is C-secure. It is important to note that the concept of C-security does not rely on quasi-concavity or continuity of profit functions. With our characterization of best response correspondences at hand, we show that a capacity and price competition game fulfills the conditions of

McLennan et al.'s result, and thus there exists a PNE. As our second main result, we show that the PNE is essentially unique (Theorem 4.5.7). To prove this, we follow an approach used in a paper of Johari, Weintraub and Van Roy [75]. However, since our model includes price caps, we need to adjust their approach in order to work for our model. In particular, the set of firms having positive capacity needs to be decomposed, where the decomposition is related to the property whether the price of a firm is equal to its cap, or strictly smaller. We finally study the worst case efficiency of the unique PNE compared to a natural benchmark, in which we relax the equilibrium conditions of the firms, but not the equilibrium conditions of the customers. We show that the unique PNE might be arbitrarily inefficient (Theorem 4.6.1), by presenting a family of instances such that the quality of the unique PNE gets arbitrarily bad.

4.1.3 Discussion of the Model and Related Work

In this subsection, we briefly discuss the used model of capacity and price competition games, as well as related work in this context.

One of the most similar models compared to ours is used in a landmark paper of Johari, Weintraub and Van Roy [75] (or in follow-up papers like [85]). Johari, Weintraub and Van Roy (JWVR for short) studied the fundamental question of existence, uniqueness and worst-case quality of PNE for capacity and price competition games which are very similar to the ones introduced by us. In difference to our model, they do not allow for price caps, but on the other hand, fairly general congestion cost functions are used. Furthermore, they mostly consider the case of *elastic* demand. The elasticity is modeled by a smooth and strictly decreasing inverse demand function mapping each volume of customers to the combined cost (congestion and price) under which the given volume of customers is participating in the game. (Depending on the generality of the allowed congestion cost functions, further concavity assumptions on the inverse demand (or demand) function are imposed.) JWVR derived several existence, uniqueness and quality results. Specifically, they showed that for models with elastic demand (and some concavity assumptions), a PNE exists and is essentially unique. The existence result uses Kakutani's fixed point theorem, where they crucially exploit the assumption that demand is elastic (this way, the best response correspondences of firms always have non-empty values). In this thesis, we focus on the case of *inelastic* demand. For this case, JWVR derived existence of PNE assuming *homogeneous* firms (all firms have the same congestion cost function) together with some further assumptions on the congestion costs and the number of firms. As shown by JWVR, homogeneity (together with some assumptions on the congestion costs) implies that there is only one symmetric equilibrium candidate profile. For this specific symmetric strategy profile, they directly proved stability using concavity arguments. This proof technique is clearly not applicable in the general non-homogeneous case (with inelastic demand). Besides the mentioned result of JWVR for homogeneous firms, not much is known in terms of existence and uniqueness of PNE for models with inelastic demand. On the other hand, many works in the transportation science and algorithmic game theory community (see, e.g., [15, 35, 36, 45, 106] and [4, 31, 39, 99], respectively) assume inelastic demand, and this case is usually considered as fundamental base case. As we show in this thesis, in terms of equilibrium existence, the case of inelastic demand is much more complicated compared to the seemingly more general case of elastic demands: best responses do not always exist in the inelastic case, putting standard fixed-point approaches out of reach. Besides this theoretical aspect, the case of inelastic demand is also interesting from a practical point of view. Litman [84] discusses various factors influencing travel demand, and summarizes several transport elasticity studies. Not surprising, for example the availability and quality of alternatives, the reason for travel, or the type of traveller, impact the elasticity. If there are few alternatives, as it is for example the case in Scandinavian freight transport, where rail and ship-based freight networks are sparse compared to road networks, demand tends to be inelastic [96]. Further examples where demand tends to be less elastic are higher value travel (as business or commute travel, in particular "urban peak-period trips"), or travel of people having higher income [84].

Further models which are related to ours are used in the papers of Harks et al. [66] and Correa et al. [32]. They consider models where firms do not choose capacities, but only prices, and the prices are upper-bounded by caps (equal for all firms in [66], firm-specific in [32]). As reported in these papers, price caps appear naturally if there are legislative regulations imposing a hard price cap in the market. In [32], it is stated that even *different* price caps for different firms is current practice in the highway market of Santiago de Chile, where 12 different operators set tolls on different urban highways, each with a specific price cap. The mentioned two papers consider the problem of a system designer who is allowed to *choose* the cap(s) in order to minimize total travel time. In particular it is observed that for firm-specific caps, the optimal flow can be induced, whereas for equal caps, the flow can be arbitrarily inefficient.

Acemoglu and Ozdaglar [2, 3] and Acemoglu et al. [1] studied models where the customers have a *reservation utility* for travelling, that is, they decide not to travel if the total cost exceeds the reservation utility. In [2] and [3], only prices are chosen by the firms, and several bounds on the worst-case efficiency of equilibria are derived. Prices *and* capacities are studied in [1], where it is assumed that the capacities represent "hard" capacities bounding the admissible customer volume for a firm. They observed that PNE do not exist if firms simultaneously choose capacities and prices. Thus, they studied a two-stage model, in which the firms first determine capacities, and in a second stage set prices. For this model, existence and worst-case efficiency of PNE is investigated.

In a very recent work, Schmand et al. [100] consider a model in which firms choose capacities for each edge of a series-parallel network, but do not directly set prices. They derived existence, uniqueness and quality results of equilibria for different assumptions on the demand.

Note that almost all of the mentioned papers on capacity and/or price competition use networks consisting of parallel edges, or slight generalizations like parallel paths [3] or series-parallel networks [100].

Finally, we want to mention that there is also a lot of work in which capacities or prices are determined *centrally* in order to reduce the total travel time of the resulting Wardrop equilibria (plus investments for the case of capacities), see for example [48, 87] for setting capacities, and [13, 45] for setting prices.

4.2   Continuity of Profit Functions

In this section, we prove a fundamental result about the continuity of the profit functions. We will use Theorem 4.2.1, which completely characterizes the strategy profiles s having the property that all profit functions Π_i, $i \in N$, are continuous at s, several times during the rest of the chapter.

Theorem 4.2.1   *Let* $s = (z, p) \in S$. *The profit function* Π_i *is continuous at* $s = (z, p)$ *for all* $i \in N$ *if and only if* $z \neq 0$ *or* $(z, p) = (0, 0)$.

Proof   We start with the profile $s = (z, p) = (0, 0)$ and show that for each firm i, her profit function Π_i is continuous at s. Let $i \in N$. For $\varepsilon > 0$, define $\delta := \min\{\frac{\varepsilon}{2\gamma_i}, \frac{\varepsilon}{2}\} > 0$, and let $s' \in S$ with $||s'|| = ||s' - s|| < \delta$ (where $||\cdot||$ denotes the Euclidean norm). If $z_i' = 0$, then $\Pi_i(s') = \Pi_i(s) = 0$. Otherwise,

$$|\Pi_i(s') - \Pi_i(s)| = |x_i(s')p_i' - \gamma_i z_i'| \leq x_i(s')p_i' + \gamma_i z_i' \leq p_i' + \gamma_i z_i'$$
$$\leq ||s'|| + \gamma_i ||s'|| < \delta(1 + \gamma_i) \leq \varepsilon$$

holds, showing that Π_i is continuous at s.

Now consider $s = (z, p) \in S$ with $z \neq 0$. We again need to show that all profit functions Π_i, $i \in N$, are continuous at s. Since $z \neq 0$, we get that $N^+ :=$

$\{j \in N : z_j > 0\} \neq \emptyset$. Furthermore, for $\delta_1 > 0$ sufficiently small, $N^+ \subseteq \{j \in N : z'_j > 0\} =: N^+(s')$ holds for all $s' = (z', p') \in S$ with $||s - s'|| < \delta_1$. Write $s' = (s'_1, s'_2) \in S$, where s'_1 denotes the strategies of the firms in N^+, and s'_2 denotes the strategies of the firms in $N \setminus N^+$. Now let $i \in N$. We need to show that Π_i is continuous at $s = (z, p)$. For all $s' \in S$ with $||s - s'|| < \delta_1$, firm i's profit is $\Pi_i(s') = x_i(s')p'_i - \gamma_i z'_i$. Thus it is sufficient to show that x_i is continuous at $s = (s_1, s_2)$. The idea of the proof is to show that, for a slightly perturbed strategy profile (s'_1, s'_2), the difference between $x_i(s)$ and $x_i(s'_1, s_2)$, as well as the difference between $x_i(s'_1, s_2)$ and $x_i(s'_1, s'_2)$, is small. In the following, let $s' \in S$ with $||s - s'|| < \delta_1$. It is well known ([13], compare also [33]) that $x(s')$ is the unique optimal solution of the following optimization problem $Q = Q(s')$:

$$\text{(Q)} \quad \min \sum_{j \in N} \int_0^{x_j} \left(\ell_j(t, z'_j) + p'_j \right) dt \quad \text{s.t.} \quad \sum_{j \in N} x_j = 1, \ x_j \geq 0 \ \forall j \in N.$$

Furthermore, $x_j(s') = 0$ for $j \notin N^+(s')$. Therefore, the values $(x_j(s'))_{j \in N^+(s')}$ are the unique optimal solution of

$$\min \sum_{j \in N^+(s')} \int_0^{x_j} \left(\frac{a_j}{z'_j}t + b_j + p'_j \right) dt \quad \text{s.t.} \quad \sum_{j \in N^+(s')} x_j = 1, \ x_j \geq 0 \ \forall j \in N^+(s'),$$

which is equivalent to

$$\max - \sum_{j \in N^+(s')} \left(\frac{a_j}{2z'_j}x_j^2 + (b_j + p'_j)x_j \right) \quad \text{s.t.} \quad \sum_{j \in N^+(s')} x_j = 1, \ x_j \geq 0 \ \forall j \in N^+(s').$$

By Berge's theorem of the maximum (Theorem 2.3.2), for all $\varepsilon > 0$ there is $0 < \delta_2 = \delta_2(\varepsilon) < \delta_1$ such that $||x(s) - x(s'_1, s_2)|| < \varepsilon$ for all $(s'_1, s_2) \in S$ with $||s - (s'_1, s_2)|| < \delta_2$. That is, x is continuous at s if we only allow changes in s_1, but not in s_2. Furthermore, if $q(s')$ denotes the optimal objective function value of $Q(s')$, and if only changes in s_1 are allowed, q is also continuous at s, i.e., for all $\varepsilon > 0$ there is $0 < \delta_3 = \delta_3(\varepsilon) < \delta_1$ such that $|q(s) - q(s'_1, s_2)| < \varepsilon$ for all $(s'_1, s_2) \in S$ with $||s - (s'_1, s_2)|| < \delta_3$. We now distinguish between the two cases that $z_i > 0$ or $z_i = 0$.

First consider the case $z_i = 0$, that is, $i \notin N^+$, and let $\varepsilon > 0$. Note that $x_i(s) = 0$, thus we need to find $\delta > 0$ such that $|x_i(s) - x_i(s')| = x_i(s') < \varepsilon$ for all $s' \in S$ with $||s - s'|| < \delta$. To this end, define

$$\delta = \delta(i, \varepsilon) := \begin{cases} \delta_3(1), & \text{if } q(s) + 1 \le b_i \varepsilon, \\ \min\{\delta_3(1), \frac{a_i \varepsilon^2}{2(q(s)+1-b_i \varepsilon)}\}, & \text{else,} \end{cases}$$

and let $s' \in S$ with $||s - s'|| < \delta$. In particular, $|z_i - z_i'| = z_i' < \delta$. Furthermore, $q(s') \le q(s_1', s_2) \le q(s) + 1$ holds since $||s - (s_1', s_2)|| \le ||s' - s|| < \delta \le \delta_3(1)$. If $z_i' = 0$, we immediately get $x_i(s') = 0 < \varepsilon$. Thus assume $z_i' > 0$ and assume, by contradiction, that $x_i(s') \ge \varepsilon$. Then, by definition of δ, we get the following contradiction:

$$q(s) + 1 \ge q(s') \ge \frac{a_i}{2z_i'} x_i(s')^2 + (b_i + p_i')x_i(s') \ge \frac{a_i}{2z_i'} \varepsilon^2 + b_i \varepsilon > \frac{a_i}{2\delta} \varepsilon^2 + b_i \varepsilon \ge q(s) + 1.$$

Therefore, $x_i(s') < \varepsilon$ holds, showing that x_i is continuous at s if $i \notin N^+$.

Now consider the case $i \in N^+$, i.e., $z_i > 0$. For $\varepsilon > 0$, we need to find $\delta > 0$ such that $|x_i(s) - x_i(s')| < \varepsilon$ for all $s' \in S$ with $||s - s'|| < \delta$. To this end, define

$$\delta := \min\{\min\{\delta(j, \frac{\varepsilon}{2n}) : j \notin N^+\}, \delta_2(\frac{\varepsilon}{2})\}$$

and let $s' \in S$ with $||s - s'|| < \delta$. In particular, $||s - (s_1', s_2)|| < \delta \le \delta_2(\frac{\varepsilon}{2})$, thus $|x_i(s) - x_i(s_1', s_2)| < \varepsilon/2$. Furthermore, since $\delta \le \delta(j, \varepsilon/(2n))$, we get $x_j(s') \le \frac{\varepsilon}{2n}$ for all $j \notin N^+$. If $x_j(s') = 0$ for all $j \notin N^+$, we get $x_i(s') = x_i(s_1', s_2)$ and thus $|x_i(s) - x_i(s')| = |x_i(s) - x_i(s_1', s_2)| < \varepsilon/2 < \varepsilon$, as desired. Otherwise, there is $j \notin N^+$ with $0 < x_j(s') \le \frac{\varepsilon}{2n}$. In particular, $z_j' > 0$. We now use a result about the sensitivity of Wardrop flows due to Englert et al. [38, Theorem 2]. They show that if the customers are not able to choose firm j's edge anymore (we say that firm j is deleted from the customers' game), the resulting change in the Wardrop flow can be bounded by the flow that j received. More formally, if $x \in [0, 1]^n$ is the Wardrop flow for the game with firms N, and $x' \in [0, 1]^{n-1}$ is the Wardrop flow if firm j is deleted from the customers' game, then $|x_k - x_k'| < x_j$ for all $k \in N \setminus \{j\}$. Obviously, changing firm j's capacity from $z_j' > 0$ to $z_j = 0$ has the same effect on the Wardrop flow as deleting firm j. Therefore, if we change, one after another, the capacities of all firms $j \notin N^+$ having $z_j' > 0$ to $z_j = 0$, we get (note that the flow values for $j \notin N^+$ are always upper-bounded by $\varepsilon/(2n)$ due to our choice of δ and the analysis of the case $z_i = 0$):

$$|x_i(s') - x_i(s_1', s_2)| \le (n - 1)\frac{\varepsilon}{2n} < \frac{\varepsilon}{2}.$$

Using this, we now get the desired inequality:

$$|x_i(s) - x_i(s')| \leq |x_i(s') - x_i(s'_1, s_2)| + |x_i(s) - x_i(s'_1, s_2)| < \frac{\varepsilon}{2} + \frac{\varepsilon}{2} = \varepsilon.$$

Altogether we showed that all profit functions Π_i are continuous at $s = (z, p)$ if $z \neq 0$.

To complete the proof, it remains to show that if all profit functions Π_i, $i \in N$, are continuous at $s = (z, p)$, then $z \neq 0$ or $(z, p) = (0, 0)$ holds. We show this by contraposition, thus assume that $s = (z, p)$ fulfills $z = 0$ and $p \neq 0$. We need to show that there is a firm i such that Π_i is not continuous at s. To this end, let $i \in N$ with $p_i > 0$. Define the sequence of strategy profiles s^n by $(z_j^n, p_j^n) := (z_j, p_j)$ for all $j \neq i$, and $(z_i^n, p_i^n) := (\frac{1}{n}, p_i)$. Obviously, $s^n \to s$ for $n \to \infty$. But for the profits, we get

$$\Pi_i(s^n) = x_i(s^n)p_i^n - \gamma_i z_i^n = p_i - \frac{\gamma_i}{n} \to_{n \to \infty} p_i > 0.$$

Since $\Pi_i(s) = 0$, this shows that Π_i is not continuous at s. $\qquad\square$

4.3 Characterization of Best Responses

The aim of this section is to derive a complete characterization of the best responses of the firms. We will make use of this characterization in all our main results, i.e., existence, uniqueness and quality of PNE. Given a firm $i \in N$ and fixed strategies $s_{-i} = (z_{-i}, p_{-i}) \in S_{-i}$ for the other firms, we characterize the set $\mathrm{BR}_i(s_{-i}) = \{s_i \in S_i : s_i$ maximizes $\Pi_i(\cdot, s_{-i})$ over $S_i\}$ of best responses of firm i to s_{-i}. To this end, we distinguish between the two cases that $z_{-i} = 0$ (Subsection 4.3.1) and $z_{-i} \neq 0$ (Subsection 4.3.2). Subsection 4.3.3 then contains the derived complete characterization. In Subsection 4.3.4, we discuss how our results about the best responses influence the applicability of Kakutani's fixed point theorem. Whenever s_{-i} is clear from the context, we simply write BR_i instead of $\mathrm{BR}_i(s_{-i})$.

4.3.1 The Case $z_{-i} = 0$

In this subsection, assume that the strategies $s_{-i} = (z_{-i}, p_{-i})$ of the other firms fulfill $z_{-i} = 0$. Under this assumption, firm i does not have a best response to s_{-i}:

Lemma 4.3.1 *If* $z_{-i} = 0$, *then* $BR_i(z_{-i}, p_{-i}) = \emptyset$.

Proof Whenever firm i chooses a strategy (z_i, p_i) with $z_i > 0$, then $x_i = 1$ holds for the induced Wardrop-flow x, thus firm i's profit is $p_i - \gamma_i z_i$. On the other hand, any strategy (z_i, p_i) with $z_i = 0$ yields a profit of 0. Thus, firm i's profit depends solely on her own strategy (z_i, p_i), and can be stated as follows:

$$\Pi_i(z_i, p_i) := \begin{cases} p_i - \gamma_i z_i, & \text{for } z_i > 0, \\ 0, & \text{for } z_i = 0. \end{cases}$$

Obviously, $\Pi_i(z_i, p_i) < C_i$ holds for each $(z_i, p_i) \in S_i$, i.e., for $z_i \geq 0$ and $0 \leq p_i \leq C_i$. On the other hand, by $(z_i, p_i) = (\varepsilon, C_i)$ for $\varepsilon > 0$, firm i gets a profit of $C_i - \gamma_i \cdot \varepsilon$ arbitrarily near to C_i, that is, $\sup\{\Pi_i(z_i, p_i) : (z_i, p_i) \in S_i\} = C_i$. This shows $\mathrm{BR}_i(z_{-i}, p_{-i}) = \emptyset$. $\qquad\square$

4.3.2 The Case $z_{-i} \neq 0$

In this subsection, assume that the strategies $s_{-i} = (z_{-i}, p_{-i})$ of the other firms fulfill $z_{-i} \neq 0$. For a strategy $s_i = (z_i, p_i)$ of firm i, write $\Pi_i(z_i, p_i) := \Pi_i(s_i, s_{-i})$ for firm i's profit function, $x(z_i, p_i) := x(s_i, s_{-i})$ for the Wardrop-flow induced by (s_i, s_{-i}) and $K(z_i, p_i) := K(s_i, s_{-i})$ for the corresponding routing cost.

For $(z_i, p_i) \in S_i$, firm i's profit is $\Pi_i(z_i, p_i) = x_i(z_i, p_i)p_i - \gamma_i z_i$. It is clear that each strategy (z_i, p_i) with $z_i = 0$ yields $x_i(z_i, p_i) = 0$, and thus $\Pi_i(z_i, p_i) = 0$. On the other hand, each strategy (z_i, p_i) with $z_i > \frac{C_i}{\gamma_i}$ yields negative profit since

$$\Pi_i(z_i, p_i) = x_i(z_i, p_i)p_i - \gamma_i z_i < 1 \cdot C_i - \gamma_i \cdot C_i/\gamma_i = 0.$$

Therefore, each best response (z_i, p_i) fulfills $z_i \leq \frac{C_i}{\gamma_i}$ since it yields nonnegative profit. Thus, BR_i can be described as the set of optimal solutions of the following optimization problem (P_i):

$$(P_i) \quad \max \; \Pi_i(z_i, p_i) \text{ subject to } z_i \in [0, C_i/\gamma_i], \; p_i \in [0, C_i].$$

(P_i) has an optimal solution, since the theorem of Weierstrass can be applied: The feasible set of (P_i) is compact and nonempty, and Π_i is continuous in (z_i, p_i) for all feasible (z_i, p_i) (see Theorem 4.2.1). Since BR_i can be described as the set of optimal solutions of (P_i), we get $\mathrm{BR}_i \neq \emptyset$.

Note that (P_i) is a *bilevel* optimization problem (since $x(z_i, p_i)$ can be described as the optimal solution of a minimization problem (see [13] and the proof of

Theorem 4.2.1 where we also used this problem)), and these problems are known to be notoriously hard to solve. The characterization of BR_i that we derive here has the advantage that it only uses *ordinary* optimization problems, namely the following two (1-dimensional) optimization problems in the variable $K \in \mathbb{R}$,

(P_i^1)

$$\max\ f_i^1(K) := \overline{x}_i(K) \cdot (K - b_i - 2\sqrt{a_i \gamma_i})$$

s.t. $2\sqrt{a_i \gamma_i} + b_i \le K$

$\quad\ K \le \sqrt{a_i \gamma_i} + b_i + C_i$

$\quad\ \overline{x}_i(K) > 0,$

(P_i^2)

$$\max\ f_i^2(K) := \overline{x}_i(K) \cdot (C_i - \frac{a_i \gamma_i}{K - b_i - C_i})$$

s.t. $\sqrt{a_i \gamma_i} + b_i + C_i < K$

$\quad\ \dfrac{a_i \gamma_i}{C_i} + b_i + C_i \le K$

$\quad\ \overline{x}_i(K) > 0,$

where

$$\overline{x}_i(K) := 1 - \sum_{j \in N(K)} \frac{(K - b_j - p_j)z_j}{a_j}$$

with $N^+ := \{j \in N \setminus \{i\} : z_j > 0\}$ and $N(K) := \{j \in N^+ : b_j + p_j < K\}$. Denote the objective functions of (P_i^1) and (P_i^2) by f_i^1 and f_i^2, respectively.

Note that $\overline{x}_i : \mathbb{R} \to \mathbb{R}$ is a continuous function which is equal to 1 for $K \le \min\{b_j + p_j : j \in N^+\}$, and strictly decreasing for $K \ge \min\{b_j + p_j : j \in N^+\} \ge 0$ with $\lim_{K \to \infty} \overline{x}_i(K) = -\infty$. Therefore, there is a unique constant $K_i^{\max} > 0$ with the property $\overline{x}_i(K_i^{\max}) = 0$. Obviously, $\overline{x}_i(K) > 0$ if and only if $K < K_i^{\max}$. Furthermore, the function $\overline{x}_i$ is closely related to Wardrop-flows, as described in the following lemma:

Lemma 4.3.2

1. If $(z_i, p_i) \in S_i$ with $x_i(z_i, p_i) > 0$, then $\overline{x}_i(K) = x_i(z_i, p_i)$ for $K := K(z_i, p_i)$.
2. If $K \ge 0$ with $\overline{x}_i(K) > 0$, and $(z_i, p_i) \in S_i$ fulfills $z_i > 0$ and $\frac{a_i}{z_i}\overline{x}_i(K) + b_i + p_i = K$, then $x_i(z_i, p_i) = \overline{x}_i(K)$ and $K(z_i, p_i) = K$.

Proof We start with statement 1. of the lemma, so let $(z_i, p_i) \in S_i$ with $x_i(z_i, p_i) > 0$. By definition of $K(z_i, p_i) =: K$, we get $\ell_j(x_j(z_i, p_i), z_j) + p_j = K$ for all $j \in N$ with $x_j(z_i, p_i) > 0$, and $\ell_j(x_j(z_i, p_i), z_j) + p_j \ge K$ for all $j \in N$ with $x_j(z_i, p_i) = 0$. Since

$$\ell_j(x_j(z_i, p_i), z_j) + p_j = \begin{cases} \dfrac{a_j}{z_j}x_j(z_i, p_i) + b_j + p_j, & \text{for } j \in N \text{ with } z_j > 0, \\ \infty, & \text{for } j \in N \text{ with } z_j = 0, \end{cases}$$

we get that $\{j \in N : x_j(z_i, p_i) > 0\} = \{j \in N^+ : b_j + p_j < K\} \cup \{i\} = N(K) \cup \{i\}$. Therefore, for each $j \in N(K)$, we get $\frac{a_j}{z_j}x_j(z_i, p_i) + b_j + p_j = K$, which is equivalent to $x_j(z_i, p_i) = \frac{(K-b_j-p_j)z_j}{a_j}$. Using $\sum_{j \in N} x_j(z_i, p_i) = 1$ yields

$$x_i(z_i, p_i) = 1 - \sum_{j \in N \setminus \{i\}} x_j(z_i, p_i) = 1 - \sum_{j \in N(K)} \frac{(K - b_j - p_j)z_j}{a_j} = \overline{x}_i(K),$$

as desired.

Now we turn to statement 2. of the lemma, so let $K \geq 0$ with $\overline{x}_i(K) > 0$ and let $(z_i, p_i) \in S_i$ be a strategy with $z_i > 0$ and $\frac{a_i}{z_i}\overline{x}_i(K) + b_i + p_i = K$. Consider the vector $x \in [0, 1]^n$ defined by

$$x_j := \begin{cases} \overline{x}_i(K), & j = i, \\ \frac{(K-b_j-p_j)z_j}{a_j}, & j \in N^+ \text{ with } b_j + p_j < K, \\ 0, & j \in N^+ \text{ with } b_j + p_j \geq K \text{ or } j \in N \setminus (N^+ \cup \{i\}). \end{cases}$$

It is clear that $x_j > 0$ holds for all $j \in N^+$ with $b_j + p_j < K$, and $x_i = \overline{x}_i(K) > 0$. Furthermore, the definition of $\overline{x}_i(K)$ yields $\sum_{j \in N} x_j = 1$. We now show that x fulfills the Wardrop equilibrium conditions. The uniqueness of the Wardrop-flow then implies $x = x(z_i, p_i)$, and $K(z_i, p_i) = K$ follows from $x_i(z_i, p_i) = \overline{x}_i(K) > 0$ and $K(z_i, p_i) = \frac{a_i}{z_i}x_i(z_i, p_i) + b_i + p_i = \frac{a_i}{z_i}\overline{x}_i(K) + b_i + p_i = K$, completing the proof. For the Wardrop equilibrium conditions, consider the effective costs of the firms:

$$c_j(x, z, p) = \begin{cases} \frac{a_i}{z_i} \cdot \overline{x}_i(K) + b_i + p_i = K, & j = i, \\ \frac{a_j}{z_j} \cdot \frac{(K-b_j-p_j)z_j}{a_j} + b_j + p_j = K, & j \in N^+ \text{ with } b_j + p_j < K, \\ b_j + p_j \geq K, & j \in N^+ \text{ with } b_j + p_j \geq K, \\ \infty > K, & j \in N \setminus (N^+ \cup \{i\}). \end{cases}$$

It is now clear that x fulfills the Wardrop equilibrium conditions. $\square$

In the following lemmata, we analyze the connection between (P_i^1) and (P_i^2) and the optimal solutions of (P_i).

Lemma 4.3.3

1. *If K is feasible for problem (P_i^1), the tuple (z_i, p_i) defined by*

$$z_i := \sqrt{a_i/\gamma_i} \cdot \overline{x}_i(K), \quad p_i := K - \sqrt{a_i\gamma_i} - b_i$$

   is feasible for (P_i), and fulfills $z_i > 0$ and $\Pi_i(z_i, p_i) = f_i^1(K)$.
2. *If K is feasible for problem (P_i^2), the tuple (z_i, p_i) defined by*

$$z_i := \frac{a_i \cdot \overline{x}_i(K)}{K - b_i - C_i}, \quad p_i := C_i$$

   is feasible for (P_i), and fulfills $z_i > 0$ and $\Pi_i(z_i, p_i) = f_i^2(K)$.

Proof We start with statement 1. of the lemma, so assume that K is feasible for problem (P_i^1). Let $z_i := \sqrt{a_i/\gamma_i} \cdot \overline{x}_i(K)$ and $p_i := K - \sqrt{a_i\gamma_i} - b_i$ as stated in 1.. The feasibility of K for (P_i^1) yields $\overline{x}_i(K) > 0$ and $2\sqrt{a_i\gamma_i} + b_i \leq K \leq \sqrt{a_i\gamma_i} + b_i + C_i$. From this we conclude $z_i > 0$ and $0 < p_i = K - \sqrt{a_i\gamma_i} - b_i \leq C_i$, thus $(z_i, p_i) \in \mathcal{S}_i$. Furthermore,

$$\frac{a_i}{z_i} \cdot \overline{x}_i(K) + b_i + p_i = \frac{a_i}{\sqrt{a_i/\gamma_i} \cdot \overline{x}_i(K)} \cdot \overline{x}_i(K) + b_i + K - \sqrt{a_i\gamma_i} - b_i = K$$

holds, thus we get $x_i(z_i, p_i) = \overline{x}_i(K)$ from statement 2. of Lemma 4.3.2. Using this, we can now show that firm i's profit for (z_i, p_i) equals the objective function value of K for (P_i^1):

$$\Pi_i(z_i, p_i) = p_i x_i(z_i, p_i) - \gamma_i z_i = (K - \sqrt{a_i\gamma_i} - b_i) \cdot \overline{x}_i(K) - \gamma_i \cdot \sqrt{a_i/\gamma_i} \cdot \overline{x}_i(K)$$
$$= \overline{x}_i(K) \cdot (K - 2\sqrt{a_i\gamma_i} - b_i) = f_i^1(K)$$

Note that the feasibility of K for (P_i^1) yields $f_i^1(K) \geq 0$. It remains to show that (z_i, p_i) is feasible for (P_i). We already know that $z_i > 0$ and $0 < p_i \leq C_i$ holds. The remaining inequality $z_i \leq \frac{C_i}{\gamma_i}$ follows from the nonnegativity of $\Pi_i(z_i, p_i) = f_i^1(K) \geq 0$ and the fact that any strategy with $z_i > \frac{C_i}{\gamma_i}$ yields negative profit for firm i.

Now turn to statement 2. of the lemma. Assume that K is feasible for (P_i^2), and let $z_i := \frac{a_i \cdot \overline{x}_i(K)}{K - b_i - C_i}$ and $p_i := C_i$. The feasibility of K for (P_i^2) implies $K > \sqrt{a_i\gamma_i} + b_i + C_i > b_i + C_i$ and $\overline{x}_i(K) > 0$, thus $z_i > 0$ holds and this yields

$(z_i, p_i) \in S_i$. Furthermore,

$$\frac{a_i}{z_i} \cdot \overline{x}_i(K) + b_i + p_i = \frac{a_i}{\frac{a_i \cdot \overline{x}_i(K)}{K - b_i - C_i}} \cdot \overline{x}_i(K) + b_i + C_i = K$$

holds, thus we get $x_i(z_i, p_i) = \overline{x}_i(K)$ from statement 2. of Lemma 4.3.2. The profit of firm i thus becomes

$$\Pi_i(z_i, p_i) = p_i x_i(z_i, p_i) - \gamma_i z_i = C_i \cdot \overline{x}_i(K) - \gamma_i \cdot \frac{a_i \overline{x}_i(K)}{K - b_i - C_i}$$

$$= \overline{x}_i(K) \cdot (C_i - \frac{a_i \gamma_i}{K - b_i - C_i}) = f_i^2(K).$$

Note that $f_i^2(K) \geq 0$ holds due to the feasibility of K for (P_i^2). As in the proof of statement 1., this implies $z_i \leq C_i/\gamma_i$, and thus (z_i, p_i) is feasible for (P_i), which completes the proof. □

In particular, Lemma 4.3.3 shows that any optimal solution of (P_i^1) or (P_i^2) yields a feasible strategy for (P_i) with the same objective function value. The next lemma shows that for certain optimal solutions of (P_i), the converse is also true.

Lemma 4.3.4 *Let (z_i^*, p_i^*) be an optimal solution of (P_i) and $K^* := K(z_i^*, p_i^*)$. If $z_i^* > 0$, then exactly one of the following two cases holds:*

1. $(z_i^*, p_i^*) = (\sqrt{a_i/\gamma_i} \cdot \overline{x}_i(K^*), K^* - \sqrt{a_i \gamma_i} - b_i)$; K^* is optimal for (P_i^1) with $f_i^1(K^*) = \Pi_i(z_i^*, p_i^*)$.
2. $(z_i^*, p_i^*) = \left(\frac{a_i \cdot \overline{x}_i(K^*)}{K^* - b_i - C_i}, C_i\right)$; K^* is optimal for (P_i^2) with $f_i^2(K^*) = \Pi_i(z_i^*, p_i^*)$.

Proof Let (z_i^*, p_i^*) with $z_i^* > 0$ and K^* as in the lemma statement, and define $x^* := x(z_i^*, p_i^*)$. Since $\Pi_i(z_i^*, p_i^*) \geq 0$ and $z_i^* > 0$ holds, $0 \leq \Pi_i(z_i^*, p_i^*) = p_i^* x_i^* - \gamma_i z_i^* < p_i^* x_i^*$ follows, which implies $x_i^* > 0$ and $p_i^* > 0$. Therefore $K^* = \frac{a_i}{z_i^*} \cdot x_i^* + b_i + p_i^*$ holds and (z_i^*, p_i^*) is an optimal solution for the following optimization problem (with variables z_i and p_i):

$$\text{(P)} \qquad \max \quad p_i x_i^* - \gamma_i z_i$$

$$\text{subject to} \quad \frac{a_i}{z_i} x_i^* + b_i + p_i = K^*$$

$$0 < z_i, \ 0 < p_i \leq C_i.$$

Note that the optimal solutions of (P) correspond to all best responses for firm i such that x^* remains the Wardrop flow. Reformulating the equality constraint in (P) yields $p_i = K^* - \frac{a_i}{z_i} x_i^* - b_i$. The constraints $0 < p_i \leq C_i$ then become (note that $K^* > b_i + p_i^* \geq b_i$ holds)

$$0 < K^* - \frac{a_i}{z_i} x_i^* - b_i \Leftrightarrow \frac{1}{z_i} < \frac{K^* - b_i}{a_i x_i^*} \Leftrightarrow z_i > \frac{a_i x_i^*}{K^* - b_i}$$

and

$$K^* - \frac{a_i}{z_i} x_i^* - b_i \leq C_i \Leftrightarrow \frac{1}{z_i} \geq \frac{K^* - b_i - C_i}{a_i x_i^*}.$$

Thus (P) is equivalent to the following problem (with variable z_i; note that $\frac{a_i x_i^*}{K^* - b_i} > 0$ holds):

$$(\mathrm{P}') \qquad \max \quad \left(K^* - \frac{a_i}{z_i} x_i^* - b_i\right) \cdot x_i^* - \gamma_i z_i$$

$$\text{subject to} \quad \frac{a_i x_i^*}{K^* - b_i} < z_i, \quad \frac{K^* - b_i - C_i}{a_i x_i^*} \leq \frac{1}{z_i}.$$

Let f be the objective function of (P$'$) and consider the derivative $f'(z_i) = \frac{a_i (x_i^*)^2}{z_i^2} - \gamma_i$. We get that f is strictly increasing for $0 < z_i < \sqrt{a_i/\gamma_i} \cdot x_i^*$ and strictly decreasing for $z_i > \sqrt{a_i/\gamma_i} \cdot x_i^*$. We now distinguish between the two cases that $z_i = \sqrt{a_i/\gamma_i} \cdot x_i^*$ is feasible for (P$'$), or not. As we will see, the former case leads to statement 1. of the lemma, and the latter case to statement 2. Note that in either case, $\overline{x}_i(K^*) = x_i^*$ holds (by statement 1. of Lemma 4.3.2).

If $z_i = \sqrt{a_i/\gamma_i} \cdot x_i^*$ is feasible for (P$'$), it is the unique optimal solution of (P$'$). But since z_i^* is also optimal for (P$'$), we get

$$z_i^* = \sqrt{a_i/\gamma_i} \cdot \overline{x}_i(K^*) \quad \text{and} \quad p_i^* = K^* - \frac{a_i}{z_i^*} \cdot \overline{x}_i(K^*) - b_i = K^* - \sqrt{a_i \gamma_i} - b_i.$$

For the profit of firm i, we get

$$\Pi_i(z_i^*, p_i^*) = p_i^* x_i^* - \gamma_i z_i^* = (K^* - \sqrt{a_i \gamma_i} - b_i) \cdot \overline{x}_i(K^*) - \gamma_i \cdot \sqrt{\frac{a_i}{\gamma_i}} \cdot \overline{x}_i(K^*)$$

$$= \overline{x}_i(K^*) \cdot (K^* - 2\sqrt{a_i \gamma_i} - b_i) = f_i^1(K^*).$$

It remains to show that K^* is optimal for (P_i^1). For feasibility, we need $2\sqrt{a_i\gamma_i}+b_i \leq K^* \leq \sqrt{a_i\gamma_i} + b_i + C_i$ and $\overline{x}_i(K^*) > 0$. The last inequality follows directly from $\overline{x}_i(K^*) = x_i^* > 0$. Using this and $\overline{x}_i(K^*) \cdot (K^* - 2\sqrt{a_i\gamma_i} - b_i) = \Pi_i(z_i^*, p_i^*) \geq 0$ yields $K^* \geq 2\sqrt{a_i\gamma_i}+b_i$. Finally, the feasibility of $z_i^* = \sqrt{a_i/\gamma_i}\cdot x_i^*$ for (P') implies

$$\frac{K^* - b_i - C_i}{a_i x_i^*} \leq \frac{\sqrt{\gamma_i}}{\sqrt{a_i}x_i^*} \Leftrightarrow K^* \leq \sqrt{a_i\gamma_i} + b_i + C_i.$$

Therefore, K^* is feasible for (P_i^1). The optimality follows from Lemma 4.3.3 and the optimality of (z_i^*, p_i^*) for (P_i).

Now turn to the case that $z_i = \sqrt{a_i/\gamma_i} \cdot x_i^*$ is not feasible for (P'). Since (P') has an optimal solution (namely z_i^*), we get that $0 < \frac{a_i x_i^*}{K^*-b_i-C_i} < \sqrt{a_i/\gamma_i} \cdot x_i^*$ holds, and therefore $z_i = \frac{a_i x_i^*}{K^*-b_i-C_i}$ is the unique optimal solution for (P'). This shows

$$z_i^* = \frac{a_i \cdot \overline{x}_i(K^*)}{K^* - b_i - C_i} \text{ and } p_i^* = K^* - \frac{a_i}{z_i^*} \cdot \overline{x}_i(K^*) - b_i = C_i.$$

The profit of firm i becomes

$$\Pi_i(z_i^*, p_i^*) = C_i \cdot \overline{x}_i(K^*) - \gamma_i \cdot \frac{a_i \overline{x}_i(K^*)}{K^* - b_i - C_i} = \overline{x}_i(K^*) \cdot (C_i - \frac{a_i\gamma_i}{K^* - b_i - C_i}) = f_i^2(K^*).$$

Since $\overline{x}_i(K^*) = x_i^* > 0$ and the profit is nonnegative,

$$C_i - \frac{a_i\gamma_i}{K^* - b_i - C_i} \geq 0 \Leftrightarrow K^* \geq \frac{a_i\gamma_i}{C_i} + b_i + C_i$$

holds. Finally,

$$z_i^* = \frac{a_i x_i^*}{K^* - b_i - C_i} < \sqrt{\frac{a_i}{\gamma_i}}x_i^* \Leftrightarrow K^* > \sqrt{a_i\gamma_i} + b_i + C_i,$$

which completes the proof since we showed that K^* is a feasible solution of problem (P_i^2) (optimality follows from Lemma 4.3.3 and the optimality of (z_i^*, p_i^*) for (P_i)). $\qquad\square$

In the next lemma, we analyze the existence of optimal solutions for the problems (P_i^1) and (P_i^2), as well as properties of such solutions.

Lemma 4.3.5

1. *If* (P_i^1) *is feasible, it has a unique optimal solution.*
2. *Assume that* (P_i^2) *is feasible.*

   - *If* $C_i \le \sqrt{a_i \gamma_i}$, *then* (P_i^2) *has a unique optimal solution.*
   - *If* $C_i > \sqrt{a_i \gamma_i}$, *then* (P_i^2) *has at most one optimal solution.*
   - *If* K_2^* *is optimal for* (P_i^2), *then* $f_i^2(K_2^*) > 0$.

3. *If* K_1^* *is optimal for* (P_i^1) *and* K_2^* *is optimal for* (P_i^2), *then* $f_i^1(K_1^*) < f_i^2(K_2^*)$.

Proof We start with statement 1. of the lemma, so assume that (P_i^1) is feasible. Note that the feasible set I_1 of (P_i^1) either is of the form $I_1 = [2\sqrt{a_i \gamma_i} + b_i, \sqrt{a_i \gamma_i} + b_i + C_i]$, or $I_1 = [2\sqrt{a_i \gamma_i} + b_i, K_i^{\max})$, depending on whether $\sqrt{a_i \gamma_i} + b_i + C_i < K_i^{\max}$ holds or not. Furthermore note that the objective function

$$
\begin{aligned}
f_i^1(K) &= \bar{x}_i(K) \cdot (K - b_i - 2\sqrt{a_i \gamma_i}) \\
&= \left(1 - \sum_{j \in N^+ : b_j + p_j < K} \frac{(K - b_j - p_j)z_j}{a_j}\right) \cdot (K - b_i - 2\sqrt{a_i \gamma_i})
\end{aligned}
$$

of (P_i^1) is continuous (over $\mathbb{R}$). From this, we can conclude that (P_i^1) has *at least one* optimal solution: For $I_1 = [2\sqrt{a_i \gamma_i} + b_i, \sqrt{a_i \gamma_i} + b_i + C_i]$, this follows directly from the theorem of Weierstrass (f_i^1 is continuous and I_1 is nonempty and compact). For $I_1 = [2\sqrt{a_i \gamma_i} + b_i, K_i^{\max})$, the theorem of Weierstrass yields that f_i^1 attains its maximum over the closure of I_1, that is, over $[2\sqrt{a_i \gamma_i} + b_i, K_i^{\max}]$. But since $f_i^1(K_i^{\max}) = 0 \ (= f_i^1(2\sqrt{a_i \gamma_i} + b_i))$, and any $K \in (2\sqrt{a_i \gamma_i} + b_i, K_i^{\max})$ fulfills $f_i^1(K) > 0$, the maximum is not attained for $K = K_i^{\max}$, and we conclude that f_i^1 also attains its maximum over I_1. Thus (P_i^1) has an optimal solution for both cases. To complete the proof of statement 1. of the lemma, it remains to show that there is also *at most one* optimal solution. We prove this by showing the following monotonicity behaviour of f_i^1 over I_1: Either f_i^1 is strictly increasing over I_1, or strictly decreasing over I_1, or strictly increasing up to a unique point, and strictly decreasing afterwards. In all three cases, we obviously get the desired statement, namely that (P_i^1) has at most one optimal solution. To prove the described monotonicity behaviour, we distinguish between three cases according to the value of $\min\{b_j + p_j : j \in N^+\}$. The first case is $\sqrt{a_i \gamma_i} + b_i + C_i \le \min\{b_j + p_j : j \in N^+\}$, which implies $N(K) = \{j \in N^+ : b_j + p_j < K\} = \emptyset$ and $\bar{x}_i(K) = 1$ for all $K \in I_1$.

We conclude that $f_i^1(K) = K - b_i - 2\sqrt{a_i\gamma_i}$ is strictly increasing over I_1 (in particular, f_i^1 reaches its unique maximum over I_1 at $K = \sqrt{a_i\gamma_i} + b_i + C_i$). Next, consider the case that $\min\{b_j + p_j : j \in N^+\} < 2\sqrt{a_i\gamma_i} + b_i$. This implies that $N(K) \neq \emptyset$ for all $K \in I_1$. Note that f_i^1 is twice differentiable on any open interval where $N(K)$ is constant, and the first and second derivative of f_i^1 then are

$$(f_i^1)'(K) = (- \sum_{j \in N(K)} \frac{z_j}{a_j}) \cdot (K - b_i - 2\sqrt{a_i\gamma_i}) + 1 - \sum_{j \in N(K)} \frac{(K - b_j - p_j)z_j}{a_j}$$

$$= 1 - \sum_{j \in N(K)} \frac{(2K - b_j - p_j - b_i - 2\sqrt{a_i\gamma_i})z_j}{a_j}$$

and

$$(f_i^1)''(K) = -2 \cdot \sum_{j \in N(K)} \frac{z_j}{a_j}.$$

Since $N(K) \neq \emptyset$ for all $K \in I_1$, we conclude that for all $K \in I_1$ where $(f_i^1)''(K)$ exists, $(f_i^1)''(K) < 0$ holds. If $N(K)$ is constant on the complete interior of I_1 (that is, on $(2\sqrt{a_i\gamma_i} + b_i, \sqrt{a_i\gamma_i} + b_i + C_i)$ or $(2\sqrt{a_i\gamma_i} + b_i, K_i^{\max})$, depending on the two possible cases for I_1), the desired monotonicity behaviour of f_i^1 over I_1 follows. Otherwise, let $\alpha_1 < \alpha_2 < \cdots < \alpha_k$ denote the different values of $b_j + p_j$, $j \in N^+$ which lie in the interior of I_1. Define $\alpha_0 := 2\sqrt{a_i\gamma_i} + b_i$ and $\alpha_{k+1} := \sup I_1$ (that is, $\alpha_{k+1} = \sqrt{a_i\gamma_i} + b_i + C_i$ or $\alpha_{k+1} = K_i^{\max}$). Then, $N(K)$ is constant on the intervals $(\alpha_{\ell-1}, \alpha_\ell)$ for $\ell \in \{1, \ldots, k+1\}$. For each $\ell \in \{1, \ldots, k\}$, the set $N(K)$ increases immediately after α_ℓ, that is, $N(\alpha_\ell) \subsetneq N(\alpha_\ell + \varepsilon)$ holds for any $\varepsilon > 0$. In particular, $N(\alpha_\ell + \varepsilon) = N(\alpha_\ell) \cup \{j \in N^+ : b_j + p_j = \alpha_\ell\}$ holds for all $0 < \varepsilon \leq \alpha_{\ell+1} - \alpha_\ell$. We now show that for any $\ell \in \{1, \ldots, k\}$, the slope of f_i^1 decreases at α_ℓ, whereby we mean that

$$(f_i^1)'_+(\alpha_\ell) < (f_i^1)'_-(\alpha_\ell)$$

holds, with $(f_i^1)'_+(\alpha_\ell)$ and $(f_i^1)'_-(\alpha_\ell)$ denoting the right and left derivative of f_i^1 at α_ℓ, respectively. This implies the desired monotonicity behaviour of f_i^1 over I_1. Analyzing the left and right derivative of f_i^1 at α_ℓ yields

$$(f_i^1)'_-(\alpha_\ell) = 1 - \sum_{j \in N(\alpha_\ell)} \frac{(2\alpha_\ell - b_j - p_j - b_i - 2\sqrt{a_i\gamma_i})z_j}{a_j}$$

and

$$(f_i^1)_+'(\alpha_\ell) = 1 - \sum_{j \in N(\alpha_\ell) \cup \{j \in N^+ : b_j + p_j = \alpha_\ell\}} \frac{(2\alpha_\ell - b_j - p_j - b_i - 2\sqrt{a_i\gamma_i})z_j}{a_j}$$

$$= (f_i^1)_-'(\alpha_\ell) - \sum_{j \in N^+ : b_j + p_j = \alpha_\ell} \frac{(2\alpha_\ell - \alpha_\ell - b_i - 2\sqrt{a_i\gamma_i})z_j}{a_j}$$

$$= (f_i^1)_-'(\alpha_\ell) - (\alpha_\ell - b_i - 2\sqrt{a_i\gamma_i}) \cdot \sum_{j \in N^+ : b_j + p_j = \alpha_\ell} \frac{z_j}{a_j}.$$

Since α_ℓ lies in the interior of I_1, we get $\alpha_\ell > 2\sqrt{a_i\gamma_i} + b_i$ and therefore the desired inequality $(f_i^1)_+'(\alpha_\ell) < (f_i^1)_-'(\alpha_\ell)$, completing the proof for the case $\min\{b_j + p_j : j \in N^+\} < 2\sqrt{a_i\gamma_i} + b_i$. The remaining case is $2\sqrt{a_i\gamma_i} + b_i \leq \min\{b_j + p_j : j \in N^+\} < \sqrt{a_i\gamma_i} + b_i + C_i$, which implies that $N(K) = \emptyset$ holds in I_1 for $K \leq \min\{b_j + p_j : j \in N^+\}$, and $N(K) \neq \emptyset$ holds in I_1 for $K > \min\{b_j + p_j : j \in N^+\}$. We can now obviously combine the arguments of the other two cases to obtain the desired monotonicity behaviour of f_i^1 also for this case. This completes the proof of statement 1. of the lemma.

Now we turn to statement 2., thus we assume that (P_i^2) is feasible. Let I_2 denote the feasible set of (P_i^2). Then, either $I_2 = [\frac{a_i\gamma_i}{C_i} + b_i + C_i, K_i^{\max})$ or $I_2 = (\sqrt{a_i\gamma_i} + b_i + C_i, K_i^{\max})$ holds, depending on whether $C_i < \sqrt{a_i\gamma_i}$ holds or not. But in both cases, I_2 is an interval with *positive* length, so that there exists $K \in I_2$ with $K > \frac{a_i\gamma_i}{C_i} + b_i + C_i$, which implies $f_i^2(K) > 0$. Therefore, if (P_i^2) has an optimal solution K_2^*, we get $f_i^2(K_2^*) > 0$. Furthermore note that the objective function

$$f_i^2(K) = \bar{x}_i(K) \cdot (C_i - \frac{a_i\gamma_i}{K - b_i - C_i})$$

$$= \left(1 - \sum_{j \in N(K)} \frac{(K - b_j - p_j)z_j}{a_j}\right) \cdot (C_i - \frac{a_i\gamma_i}{K - b_i - C_i})$$

of (P_i^2) is continuous over $(b_i + C_i, \mathbb{R})$. Using this, we can show that (P_i^2) has *at least one* optimal solution, if $C_i \leq \sqrt{a_i\gamma_i}$ holds: Due to the theorem of Weierstrass, f_i^2 attains its maximum over $[\frac{a_i\gamma_i}{C_i} + b_i + C_i, K_i^{\max}]$, the closure of I_2. But since $f_i^2(\frac{a_i\gamma_i}{C_i} + b_i + C_i) = 0 = f_i^2(K_i^{\max})$, and $f_i^2(K) > 0$ for any $K \in (\frac{a_i\gamma_i}{C_i} + b_i + C_i, K_i^{\max})$, the maximum is not attained at $K = \frac{a_i\gamma_i}{C_i} + b_i + C_i$ or $K = K_i^{\max}$, which shows that f_i^2 also attains its maximum over I_2. Thus, if $C_i \leq \sqrt{a_i\gamma_i}$ holds, (P_i^2) has at least one optimal solution. To complete the proof of statement 2., it remains to show that (P_i^2) has *at most one* optimal solution (in the general case). As in the proof of statement 1. of the lemma, we proof this by showing that f_i^2 exhibits a certain monotonicity behaviour over I_2, namely: Either f_i^2 is strictly decreasing over I_2, or strictly increasing up to a unique point, and strictly decreasing afterwards. Note that f_i^2 cannot be strictly increasing over I_2, due to the continuity of f_i^2 and the fact that $f_i^2(K_i^{\max}) = 0 < f_i^2(K)$ holds for any K in the interior of I_2. The remaining proof is very similar to the proof of statement 1.. First, f_i^2 is twice differentiable on any open interval where $N(K)$ is constant. The first and second derivative of f_i^2 then are

$$(f_i^2)'(K) = -\sum_{j \in N(K)} \frac{z_j}{a_j} \cdot \left(C_i - \frac{a_i \gamma_i}{K - b_i - C_i}\right) + \overline{x}_i(K) \cdot \frac{a_i \gamma_i}{(K - b_i - C_i)^2} \quad \text{and}$$

$$(f_i^2)''(K) = -\sum_{j \in N(K)} \frac{z_j}{a_j} \cdot \frac{a_i \gamma_i}{(K - b_i - C_i)^2} - \sum_{j \in N(K)} \frac{z_j}{a_j} \cdot \frac{a_i \gamma_i}{(K - b_i - C_i)^2}$$

$$+ \overline{x}_i(K) \cdot \frac{-2a_i \gamma_i}{(K - b_i - C_i)^3}$$

$$= -\frac{2a_i \gamma_i}{(K - b_i - C_i)^2} \cdot \left(\sum_{j \in N(K)} \frac{z_j}{a_j} + \frac{\overline{x}_i(K)}{K - b_i - C_i}\right).$$

Since $\overline{x}_i(K) > 0$ for all $K \in I_2$, we conclude that for all $K \in I_2$ where $(f_i^2)''(K)$ exists, $(f_i^2)''(K) < 0$ holds. If $N(K)$ is constant on the complete interior of I_2, the desired monotonicity behaviour of f_i^2 over I_2 follows, otherwise let $\beta_1 < \beta_2 < \cdots < \beta_k$ denote the different values of $b_j + p_j$, $j \in N^+$ which lie in the interior of I_2. We show that the slope of f_i^2 decreases at β_ℓ, i.e., $(f_i^2)'_+(\beta_\ell) < (f_i^2)'_-(\beta_\ell)$ holds, which implies the desired monotonicity behaviour of f_i^2 over I_2. Analyzing the left and right derivative yields

$$(f_i^2)'_-(\beta_\ell) = -\sum_{j \in N(\beta_\ell)} \frac{z_j}{a_j} \cdot \left(C_i - \frac{a_i \gamma_i}{\beta_\ell - b_i - C_i}\right) + \overline{x}_i(\beta_\ell) \cdot \frac{a_i \gamma_i}{(\beta_\ell - b_i - C_i)^2} \quad \text{and}$$

$$(f_i^2)'_+(\beta_\ell) = -\sum_{j \in N(\beta_\ell) \cup \{j \in N^+ : b_j + p_j = \beta_\ell\}} \frac{z_j}{a_j} \cdot \left(C_i - \frac{a_i \gamma_i}{\beta_\ell - b_i - C_i}\right) + \overline{x}_i(\beta_\ell) \cdot \frac{a_i \gamma_i}{(\beta_\ell - b_i - C_i)^2}$$

$$= (f_i^2)'_-(\beta_\ell) - \sum_{j \in N^+ : b_j + p_j = \beta_\ell} \frac{z_j}{a_j} \cdot \left(C_i - \frac{a_i \gamma_i}{\beta_\ell - b_i - C_i}\right)$$

Since β_ℓ lies in the interior of I_2, we get $C_i - \frac{a_i \gamma_i}{\beta_\ell - b_i - C_i} > 0$, and thus the desired inequality $(f_i^2)'_+(\beta_\ell) < (f_i^2)'_-(\beta_\ell)$, completing the proof of statement 2. of the lemma.

Finally we show statement 3. of the lemma, so assume that K_1^* and K_2^* are the optimal solutions of (P_i^1) and (P_i^2). Since (P_i^1) and (P_i^2) have to be feasible, $\sqrt{a_i \gamma_i} \le C_i$ and $\sqrt{a_i \gamma_i} + b_i + C_i < K_i^{\max}$ holds. This implies that the feasible set of (P_i^1) is $I_1 = [2\sqrt{a_i \gamma_i} + b_i, \sqrt{a_i \gamma_i} + b_i + C_i]$ and the feasible set of (P_i^2) is $I_2 = (\sqrt{a_i \gamma_i} + b_i + C_i, K_i^{\max})$. Let $\bar{K} := \sqrt{a_i \gamma_i} + b_i + C_i$. Then,

$$
\begin{aligned}
f_i^1(\bar{K}) &= \bar{x}_i(\sqrt{a_i \gamma_i} + b_i + C_i) \cdot (\sqrt{a_i \gamma_i} + b_i + C_i - b_i - 2\sqrt{a_i \gamma_i}) \\
&= \bar{x}_i(\sqrt{a_i \gamma_i} + b_i + C_i) \cdot (C_i - \sqrt{a_i \gamma_i}) \\
&= \bar{x}_i(\sqrt{a_i \gamma_i} + b_i + C_i) \cdot \left(C_i - \frac{a_i \gamma_i}{\sqrt{a_i \gamma_i} + b_i + C_i - b_i - C_i} \right) \\
&= f_i^2(\bar{K})
\end{aligned}
$$

holds. If additionally the slope of f_i^1 in $\bar{K}$ is greater than or equal to the slope of f_i^2 in $\bar{K}$, whereby we mean that $(f_i^1)'_-(\bar{K}) \ge (f_i^2)'_+(\bar{K})$ holds, we get $f_i^1(K_1^*) = f_i^1(\bar{K}) < f_i^2(K_2^*)$ from our analysis of f_i^1 and f_i^2 in the proofs of the statements 1. and 2. (note that f_i^2 is strictly increasing on the interval $(\sqrt{a_i \gamma_i} + b_i + C_i, K_2^*]$). The remaining inequality for the slopes follows from

$$
\begin{aligned}
(f_i^1)'_-(\bar{K}) &= 1 - \sum_{j \in N(\bar{K})} \frac{(2\bar{K} - b_j - p_j - b_i - 2\sqrt{a_i \gamma_i}) z_j}{a_j} \\
&= 1 - \sum_{j \in N(\bar{K})} \frac{(2C_i + b_i - b_j - p_j) z_j}{a_j} \quad \text{and}
\end{aligned}
$$

$$
\begin{aligned}
(f_i^2)'_+(\bar{K}) &= - \sum_{j \in N(\bar{K}) \cup \{j \in N^+ : b_j + p_j = \bar{K}\}} \frac{z_j}{a_j} \cdot (C_i - \frac{a_i \gamma_i}{\bar{K} - b_i - C_i}) + \bar{x}_i(\bar{K}) \cdot \frac{a_i \gamma_i}{(\bar{K} - b_i - C_i)^2} \\
&= - \sum_{j \in N(\bar{K}) \cup \{j \in N^+ : b_j + p_j = \bar{K}\}} \frac{z_j}{a_j} \cdot (C_i - \sqrt{a_i \gamma_i}) + \bar{x}_i(\bar{K}) \cdot 1 \\
&= - \sum_{j \in N(\bar{K})} \frac{z_j}{a_j} \cdot (C_i - \sqrt{a_i \gamma_i}) - \sum_{j \in N^+ : b_j + p_j = \bar{K}} \frac{z_j}{a_j} \cdot (C_i - \sqrt{a_i \gamma_i}) \\
&\quad + 1 - \sum_{j \in N(\bar{K})} \frac{(\bar{K} - b_j - p_j) z_j}{a_j} \\
&= 1 - \sum_{j \in N(\bar{K})} \frac{(\bar{K} - b_j - p_j + C_i - \sqrt{a_i \gamma_i}) z_j}{a_j} - \sum_{j \in N^+ : b_j + p_j = \bar{K}} \frac{z_j}{a_j} \cdot (C_i - \sqrt{a_i \gamma_i})
\end{aligned}
$$

$$= 1 - \sum_{j \in N(\tilde{K})} \frac{(2C_i + b_i - b_j - p_j)z_j}{a_j} - \sum_{j \in N^+ : b_j + p_j = \tilde{K}} \frac{z_j}{a_j} \cdot (C_i - \sqrt{a_i \gamma_i})$$

$$= (f_i^1)'_-(\tilde{K}) - \sum_{j \in N^+ : b_j + p_j = \tilde{K}} \frac{z_j}{a_j} \cdot (C_i - \sqrt{a_i \gamma_i})$$

$$\leq (f_i^1)'_-(\tilde{K}),$$

where the inequality follows from $\sqrt{a_i \gamma_i} \leq C_i$. $\square$

4.3.3   The Characterization

The following theorem provides a complete characterization of the best response correspondence. We will make use of this characterization several times during the rest of the chapter.

Theorem 4.3.6   (characterization of best responses). *For a firm $i \in N$ and fixed strategies $s_{-i} = (z_{-i}, p_{-i}) \in S_{-i}$ of the other firms, the set $BR_i = BR_i(s_{-i})$ of best responses of firm i to s_{-i} is given as indicated in the following Table 4.1, where the first column contains BR_i and the second column contains the conditions on s_{-i} under which BR_i has the stated form. For $j = 1, 2$, K_j^* denotes the unique optimal solution of problem (P_i^j), if this problem has an optimal solution.*

Furthermore, if $BR_i(s_{-i})$ consists of a unique best response $s_i = (z_i, p_i)$ of firm i to s_{-i}, we get $z_i > 0$ and $\Pi_i(s_i, s_{-i}) > 0$.

Table 4.1 Characterization of BR_i

$\{(z_i, p_i)\} = BR_i$	conditions
$\emptyset$	$z_{-i} = 0$
$\{(0, p_i) : 0 \leq p_i \leq C_i\}$	$z_{-i} \neq 0$, (P_i^1) and (P_i^2) infeasible
$\left\{ \left(\frac{a_i \cdot \bar{x}_i(K_2^*)}{K_2^* - b_i - C_i}, C_i \right) \right\}$	$z_{-i} \neq 0$, (P_i^2) has an optimal solution
$\left\{ \left(\sqrt{\frac{a_i}{\gamma_i}} \cdot \bar{x}_i(K_1^*), K_1^* - \sqrt{a_i \gamma_i} - b_i \right) \right\}$	$z_{-i} \neq 0$, (P_i^1) feasible, (P_i^2) has no optimal solution

Proof Note that if $BR_i(s_{-i})$ consists of a *unique* best response $s_i = (z_i, p_i)$ of firm i to s_{-i}, then $z_i > 0$ and $\Pi_i(s_i, s_{-i}) > 0$ hold: Otherwise, any strategy $(z_i', p_i') \in \{0\} \times [0, C_i]$ is a best response, too, contradicting the uniqueness assumption.

Now turn to the proof of the characterization. We show that the case distinction covers all possible cases, and that the given representation for BR_i is correct for each case. If $z_{-i} = 0$, Lemma 4.3.1 shows $\mathrm{BR}_i = \emptyset$. For the rest of the proof, assume $z_{-i} \neq 0$. Then, firm i has at least one best response to s_{-i}, since BR_i can be described as the set of optimal solution of the problem (P_i), and this problem has an optimal solution (as shown in the beginning of Subsection 4.3.2). If (P_i^1) and (P_i^2) are both infeasible, Lemma 4.3.4 implies that each best response (z_i, p_i) fulfills $z_i = 0$. Therefore, $\mathrm{BR}_i = \{0\} \times [0, C_i]$. For the remaining proof, assume that at least one of (P_i^1) and (P_i^2) is feasible.

First consider the case that (P_i^2) has an optimal solution. It follows from 2. of Lemma 4.3.5 that the solution is unique, and, if K_2^* denotes this unique solution, that $f_i^2(K_2^*) > 0$. Now let (z_i, p_i) be an arbitrary best response of player i to s_{-i}. We need to show that $(z_i, p_i) = (\frac{a_i \bar{x}_i(K_2^*)}{K_2^* - b_i - C_i}, C_i)$ holds. First, statement 2. of Lemma 4.3.3 shows that $\Pi_i(z_i, p_i) \geq f_i^2(K_2^*) > 0$. Thus $z_i > 0$ holds, since $z_i = 0$ yields a profit of 0. Note that either (P_i^1) is infeasible, or it has a unique optimal solution K_1^* with $f_i^1(K_1^*) < f_i^2(K_2^*) \leq \Pi_i(z_i, p_i)$ (see 1. and 3. of Lemma 4.3.5). In both cases, Lemma 4.3.4 yields $(z_i, p_i) = (\frac{a_i \bar{x}_i(K_2^*)}{K_2^* - b_i - C_i}, C_i)$, as desired.

Now assume that (P_i^2) does not have an optimal solution (either (P_i^2) is infeasible, or it is feasible, but the maximum is not attained). We first show that (P_i^1) is feasible. If (P_i^2) is infeasible, (P_i^1) is feasible since we assumed that at least one of the two problems is feasible. Otherwise (P_i^2) is feasible, but does not have an optimal solution. Then, $C_i > \sqrt{a_i \gamma_i}$ follows from 2. of Lemma 4.3.5, and $\sqrt{a_i \gamma_i} + b_i + C_i < K_i^{\max}$ follows since (P_i^2) is feasible. Together, $2\sqrt{a_i \gamma_i} + b_i < \sqrt{a_i \gamma_i} + b_i + C_i < K_i^{\max}$ holds, showing that (P_i^1) is feasible. By 1. of Lemma 4.3.5 we then get that (P_i^1) has a unique optimal solution K_1^*. Furthermore, each best response (z_i, p_i) with $z_i > 0$ fulfills $(z_i, p_i) = (\sqrt{a_i/\gamma_i} \cdot \bar{x}_i(K_1^*), K_1^* - \sqrt{a_i \gamma_i} - b_i)$ (see Lemma 4.3.4). To complete the proof, we need to show that there is no best response (z_i, p_i) with $z_i = 0$. This follows from Lemma 4.3.3 if $f_i^1(K_1^*) > 0$. Thus it remains to show that $f_i^1(K_1^*) > 0$ holds. Assume, by contradiction, that $f_i^1(K_1^*) = 0$. This implies that $K_1^* = 2\sqrt{a_i \gamma_i} + b_i$ is the only feasible solution for (P_i^1), which in turn yields $\sqrt{a_i \gamma_i} = C_i$ and $\sqrt{a_i \gamma_i} + b_i + C_i < K_i^{\max}$. But this implies that (P_i^2) is feasible and has an optimal solution (by 2. of Lemma 4.3.5), contradicting our assumption that (P_i^2) does not have an optimal solution. $\qquad\square$

4.3.4 Discussion

We now briefly discuss consequences of the characterization of the best response correspondences with respect to applying Kakutani's fixed point theorem (Theorem 2.3.1). His theorem in particular requires that for each firm i and each vector $s_{-i} = (z_{-i}, p_{-i})$ of strategies of the other firms, the set $\mathrm{BR}_i(s_{-i})$ of best responses is nonempty and convex. But as we have seen in Lemma 4.3.1, the set $\mathrm{BR}_i(s_{-i})$ can be empty, namely if $z_{-i} = 0$. On the other hand, a profile with $z_{-i} = 0$ for some firm i will of course never be a PNE.

A first natural approach to overcome the problem of empty best responses is the following. Given a strategy profile $s = (z, p)$ such that $z_{-i} = 0$ for some firm i, redefine, for each such firm i, the set $\mathrm{BR}_i(s_{-i})$ by some *suitable* nonempty convex set. "Suitable" here means that the best response correspondence $\mathrm{BR}_i : \mathcal{S}_{-i} \twoheadrightarrow \mathcal{S}_i$ of firm i is closed, and at the same time, s must not be a fixed point of the global best response correspondence $\mathrm{BR} : \mathcal{S} \twoheadrightarrow \mathcal{S}$. But unfortunately, these two goals are not compatible: Consider the strategy profile $s = (z, p)$ with $(z_i, p_i) = (0, C_i)$ for all firms i. To obtain that BR_i is closed, $(0, C_i) \in \mathrm{BR}_i(s_{-i})$ needs to hold for all i (note that whenever $s_{-i}^n \to_{n \to \infty} s_{-i}$ is a sequence of strategies of the other players converging to s_{-i}, and $s_i^n \in \mathrm{BR}_i(s_{-i}^n)$ for all $n \in N$, then $s_i^n \to_{n \to \infty} (0, C_i)$, since player i's profit approaches C_i). But this implies that s is a fixed point of BR.

Another intuitive idea is to consider a game in which each firm has an initial capacity of some $\varepsilon > 0$. If this game has a PNE for each ε, the limit for ε going to zero should be a PNE for our original capacity and price competition game. For the game with at least ε capacity, one can also characterize the best response correspondences (now by optimal solutions of *three* optimization problems), but the main problem is that an analogue of Lemma 4.3.5 may not hold anymore. As a consequence, it is not clear if the best responses are always convex, and we again do not know how to apply Kakutani's theorem. Instead, we show existence of PNE by using a result of McLennan et al. [89].[2]

4.4   Existence of Equilibria

In this section, we show that each capacity and price competition game has a PNE. A frequently used tool to show existence of PNE is Kakutani's fixed point theorem. But, as discussed in Subsection 4.3.4, we cannot directly apply this result to show existence of PNE. Furthermore, the existence theorem of Reny [95] can also not be used, since a capacity and price competition game is not quasiconcave in general (see Example 4.4.7). Instead, we turn to another existence result due to McLennan, Monteiro and Tourky [89]. They introduced a concept called *C-security* and they

---

[2]Perhaps interesting, McLennan et al. identify a non-trivial restriction of the players' non-equilibrium strategies so that they can eventually apply Kakutani's theorem.

showed that if the game is C-secure at each strategy profile which is not a PNE, then a PNE exists. Informally, the game is C-secure at a strategy profile s if there is a vector $\alpha \in \mathbb{R}^n$ satisfying the following two properties: First, each firm i has some securing strategy for α_i which is robust to small deviations of the other firms, i.e., firm i always achieves a profit of at least α_i by playing this strategy even if the other firms slightly deviate from their strategies in s_{-i}. The second property requires for each slightly perturbed strategy profile s' resulting from s, that there is at least one firm i such that her perturbed strategy s_i' can (in some sense) be strictly separated from all strategies achieving a profit of α_i, so in particular from all her securing strategies. One can think of firm i being "not happy" with her perturbed strategy s_i' since she could achieve a higher profit. This already indicates the connection between a strategy profile which is not a PNE, and C-security. We will see that for certain strategy profiles, α_i can be chosen as the profit that firm i gets by playing a best response to s_{-i}.[3] Then, firm i's securing strategies for α_i are related to her set of best responses, and we need to "strictly separate" these best responses from s_i. At this point, our characterization of best responses in Theorem 4.3.6 becomes useful.

We now formally describe McLennnan et al.s' result in our context. First of all, note that they consider games with compact convex strategy sets and bounded profit functions. In a capacity and price competition game, the strategy set $\mathcal{S}_i = \{(z_i, p_i) : 0 \le z_i, 0 \le p_i \le C_i\}$ of firm i is not compact a priori. But since z_i will never be larger than C_i/γ_i in any best response, and thus in any PNE (see the discussion on page 191), we can redefine $\mathcal{S}_i := \{(z_i, p_i) : 0 \le z_i \le C_i/\gamma_i, 0 \le p_i \le C_i\}$ without changing the set of PNE of the game, for any firm i. Furthermore, this also does not change the best responses, so Theorem 4.3.6 continues to hold. Using the redefined strategies, for any firm i and any strategy profile s, the profit of firm i is bounded by $-C_i \le \Pi_i(s) \le C_i$. For a strategy profile $s \in \mathcal{S}$, firm $i \in N$ and $\alpha_i \in \mathbb{R}$ let

$$B_i(s, \alpha_i) := \{s_i' \in \mathcal{S}_i : \Pi_i(s_i', s_{-i}) \ge \alpha_i\} \text{ and } C_i(s, \alpha_i) := \text{conv } B_i(s, \alpha_i),$$

where conv $B_i(s, \alpha_i)$ denotes the convex hull of $B_i(s, \alpha_i)$.

Definition 4.4.1 A firm i can *secure a profit* $\alpha_i \in \mathbb{R}$ *on* $\mathcal{S}' \subseteq \mathcal{S}$, if there is some $s_i \in \mathcal{S}_i$ such that $s_i \in B_i(s', \alpha_i)$ for all $s' \in \mathcal{S}'$. We say that firm i can *secure* α_i *at* $s \in \mathcal{S}$, if she can secure α_i on $U \cap \mathcal{S}$ for some open set U with $s \in U$.

---

[3]More precisely, we need to choose α_i slightly smaller than the profit of a best response.

Definition 4.4.2 The game is *C-secure on* $S' \subseteq S$, if there is an $\alpha \in \mathbb{R}^n$ such that the following conditions hold:

(i) Every firm i can secure α_i on S'.
(ii) For any $s' \in S'$, there exists some firm i with $s_i' \notin C_i(s', \alpha_i)$.

The game is *C-secure at* $s \in S$, if it is C-secure on $U \cap S$ for some open set U with $s \in U$.

We can now state the existence result of McLennan et al.:

Theorem 4.4.3 (Proposition 2.7 in [89]). *If the game is C-secure at each $s \in S$ that is not a PNE, then the game has a PNE.*

We now turn to capacity and price competition games and show the existence of a PNE by using Theorem 4.4.3, i.e., we show that if a given strategy profile $s = (z, p)$ is not a PNE, then the game is C-secure at s. To this end, we distinguish between the two cases that there are at least two firms i with $z_i > 0$ (Lemma 4.4.4), or not (Lemma 4.4.5). Both lemmata together then imply the desired existence result. Note that the mentioned case distinction is equivalent to the case distinction that each firm i has a best response for s_{-i}, or there is at least one firm i with $\mathrm{BR}_i(s_{-i}) = \emptyset$ (see Theorem 4.3.6).

We start with the case that all best responses exist. The proof of the following lemma follows an argument in [89, p. 1647 f.] where McLennan et al. show that Theorem 4.4.3 implies the existence result of [92].

Lemma 4.4.4 *Let $s = (z, p) \in S$ be a strategy profile which is not a PNE. Assume that there are at least two firms $i \in N$ such that $z_i > 0$ holds. Then the game is C-secure at s.*

Proof We first introduce some notation used in this proof. Let $S' \subseteq S$ be a subset of the strategy profiles and $i \in N$. By $S_i' \subseteq S_i$, we denote the projection of S' into S_i, the set of firm i's strategies, and $S_{-i}' \subseteq S_{-i}$ denotes the projection of S' into $S_{-i} = \times_{j \in N \setminus \{i\}} S_j$, the set of strategies of the other firms. Note that since z has at least two positive entries z_i, all strategy profiles $s' = (z', p')$ in a sufficiently small open neighbourhood of s also have at least two entries $z_i' > 0$. In the following, whenever we speak of an open set U containing s, we implicitly require U small enough to fulfill this property. Furthermore, since it is clear that we are only interested in the elements of U which are strategy profiles, we simply write U instead of $U \cap S$.

Consequently, $s' \in U$ denotes a strategy profile contained in U. Now we turn to the actual proof.

Since we assumed that at least two firms have positive capacity at $s = (z, p)$, we get that $z_{-i} \neq 0$ holds for each firm i. Thus, by Theorem 4.3.6, each firm i has a best response to s_{-i}, and the set of best responses either is a singleton, or consists of all strategies $(0, p_i)$ for $0 \leq p_i \leq C_i$. Since s is not a PNE, there is at least one firm j such that $s_j = (z_j, p_j)$ is not a best response, i.e., either $s_j \neq s_j^*$ for the unique best response s_j^*, or $z_j > 0$, and all best responses $s_j^* = (z_j^*, p_j^*)$ fulfill $z_j^* = 0$. In both cases it is clear that there is a hyperplane H which strictly separates s_j from the set of best responses to s_{-j}.

We now turn to the properties in Definition 4.4.2. For each firm i, let s_i^* be a best response of firm i to s_{-i} and $\beta_i := \Pi_i(s_i^*, s_{-i})$. We know from Theorem 4.2.1 that Π_i is continuous at (s_i^*, s_{-i}) for each firm i. Therefore, for each $\varepsilon > 0$, there is an open set $U(\varepsilon)$ containing s such that $\Pi_i(s_i^*, s_{-i}') \geq \beta_i - \varepsilon$ for each $s' \in U(\varepsilon)$ and each firm i. That is, each firm i can secure $\beta_i - \varepsilon$ on $U(\varepsilon)$. Now turn to the second property of Definition 4.4.2 and consider firm j. We now show that there is an $\varepsilon > 0$ and an open set $U \subseteq U(\varepsilon)$ containing s, such that for each $s' \in U$, the hyperplane H (which strictly separates s_j and $B_j(s, \beta_j)$) also strictly separates s_j' and $B_j(s', \beta_j - \varepsilon)$, thus $s_j' \notin C_j(s', \beta_j - \varepsilon)$ (see Figure 4.1 for an illustration). Since each firm i can secure $\beta_i - \varepsilon$ on $U \subseteq U(\varepsilon)$, both properties of Definition 4.4.2 are fulfilled, completing the proof.

To this end, choose an open set V containing $B_j(s, \beta_j)$ such that H strictly separates s_j and V. Since $\mathcal{S}_j \setminus V$ is a compact set and Π_j is continuous in $(\tilde{s}_j, s_{-j})$ for all $\tilde{s}_j \in \mathcal{S}_j \setminus V$ (by Theorem 4.2.1), we get that $f(s_{-j}) := \max\{\Pi_j(\tilde{s}_j, s_{-j}) : \tilde{s}_j \in \mathcal{S}_j \setminus V\}$ exists. Furthermore, since $B_j(s, \beta_j) \subset V$, we get $f(s_{-j}) < \beta_j$. Let $0 < \varepsilon < \beta_j - f(s_{-j})$, thus $f(s_{-j}) < \beta_j - \varepsilon$. Note that if we consider, for an open neighbourhood $\bar{U}$ of s and for fixed $s_{-j}' \in \bar{U}_{-j}$, the problem of maximizing $\Pi_j(\tilde{s}_j, s_{-j}')$ subject to $\tilde{s}_j \in \mathcal{S}_j \setminus V$, Berge's theorem of the maximum (Theorem 2.3.2) yields that $f(s_{-j}') := \max\{\Pi_j(\tilde{s}_j, s_{-j}') : \tilde{s}_j \in \mathcal{S}_j \setminus V\}$ is a continuous function. Using the continuity of f, there is an open set $U \subseteq U(\varepsilon)$ containing s such that $f(s_{-j}') < \beta_j - \varepsilon$ for all $s_{-j}' \in U_{-j}$. Additionally, let U be small enough that H strictly separates U_j and V. Now we have the desired properties: For each $s' \in U$ and for each $\tilde{s}_j \in \mathcal{S}_j \setminus V$, we get $\Pi_j(\tilde{s}_j, s_{-j}') < \beta_j - \varepsilon$, thus $B_j(s', \beta_j - \varepsilon) \subset V$. Since $s_j' \in U_j$ and H strictly separates U_j and V, we get that H strictly separates s_j' and $B_j(s', \beta_j - \varepsilon)$, as desired. $\square$

Fig. 4.1 Illustration of the
proof construction for the
case $B_j(s, \beta_j) = \{s_j^*\}$. Note
that it is not necessary that
$s_j^* \in B_j(s', \beta_j - \varepsilon)$

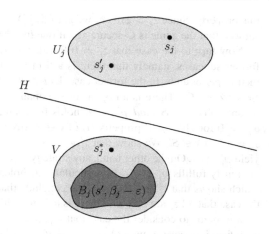

It remains to analyze the strategy profiles $s = (z, p)$ with at most one positive
z_i. Note that these profiles cannot be PNE.

Lemma 4.4.5 *Let $s = (z, p) \in S$ be a strategy profile such that $z_i > 0$ holds for
at most one firm i. Then the game is C-secure at s.*

Proof We distinguish between the two cases that there is a firm with positive
capacity, or all capacities are zero. In the former case, assume that $z_i > 0$ for
firm i, and $z_j = 0$ for all $j \neq i$ hold. Choose $\alpha_i \in (C_i - \gamma_i z_i, C_i)$ and
$0 < \varepsilon < \min\{z_i, \frac{\alpha_i + \gamma_i z_i - C_i}{\gamma_i}\}$. Then, there is an open set U containing s such
that firm i can secure α_i on $U \cap S =: S'$ (note that by choosing U sufficiently
small, firm i can secure each profit $< C_i$ on S') and $|z_i' - z_i| < \varepsilon$ holds for each
$s' = (z', p') \in S'$. For $j \neq i$, set $\alpha_j := 0$. It is clear that each firm $j \neq i$ can
secure $\alpha_j = 0$ on S' (by any strategy with zero capacity). This way, property (i)
of C-security is fulfilled. For property (ii), let $s' = (z', p') \in S'$. We show that
$s_i' \notin C_i(s', \alpha_i)$ holds. To this end, note that any strategy $s_i^* = (z_i^*, p_i^*) \in B_i(s', \alpha_i)$,
i.e., with $\Pi_i(s_i^*, s_{-i}') \geq \alpha_i$, fulfills $z_i^* \leq z_i - \varepsilon$, since, for $z_i^* > z_i - \varepsilon > 0$, we get

$$\Pi_i(s_i^*, s_{-i}') = x_i(s_i^*, s_{-i}')p_i^* - \gamma_i z_i^* \leq C_i - \gamma_i z_i^* < C_i - \gamma_i(z_i - \varepsilon) < \alpha_i,$$

where the last inequality follows from the choice of ε. Clearly, any strategy in
$C_i(s', \alpha_i)$, i.e., any convex combination of strategies in $B_i(s', \alpha_i)$, then also has

this property. Since $z_i' > z_i - \varepsilon$, we get $(z_i', p_i') \notin C_i(s', \alpha_i)$, as desired. Thus we showed that the game is C-secure at s if one firm has positive capacity.

Now turn to the case that $z_i = 0$ for all $i \in N$. We distinguish between two further subcases, namely that there is a firm i with $p_i < C_i$, or all prices are at their upper bounds. In the former case, let $i \in N$ with $p_i < C_i$, and choose α_i with $p_i < \alpha_i < C_i$. There is an open set U containing s such that firm i can secure α_i on $U \cap S =: S'$ and $p_i' < \alpha_i$ holds for each $s' = (z', p') \in S'$. By setting $\alpha_j := 0$ for all $j \neq i$, property (i) of C-security is fulfilled. For property (ii), let $s' = (z', p') \in S'$. We show that $s_i' \notin C_i(s', \alpha_i)$ holds. Our assumptions about S' yield $p_i' < \alpha_i$. On the other hand, any strategy $s_i^* = (z_i^*, p_i^*)$ with $\Pi_i(s_i^*, s_{-i}') \geq \alpha_i$ obviously fulfills $p_i^* > \alpha_i$. In particular, this holds for any strategy in $C_i(s', \alpha_i)$, which shows that $s_i' \notin C_i(s', \alpha_i)$. We conclude that the game is C-secure at s for the case that all z_i are zero *and* there is a firm i with $p_i < C_i$.

It remains to consider the case that $(z_i, p_i) = (0, C_i)$ holds for all firms i. For each firm i, choose α_i with $(1 - \frac{a_i}{2(a_i + C_i)})C_i < \alpha_i < C_i$. Note that this implies

$$\frac{1}{2} < \frac{1}{2} + \frac{C_i - \alpha_i}{a_i} < \frac{\alpha_i}{C_i}. \tag{4.1}$$

There is an open set U containing s such that each firm i can secure α_i on $U \cap S =: S'$ and $z_i' < 1$ holds for each profile $s' = (z', p') \in S'$. Thus, property (i) of C-security is fulfilled. For property (ii), let $s' = (z', p') \in S'$. In the following, $s_i^* = (z_i^*, p_i^*)$ denotes a strategy of firm i with $\Pi_i(s_i^*, s_{-i}') \geq \alpha_i > 0$. We say that s_i^* *achieves a profit of at least* α_i. Obviously, $z_i^* > 0$ and $p_i^* > \alpha_i$ hold. Furthermore, $x_i(s_i^*, s_{-i}') > \frac{1}{2}$, since $\Pi_i(s_i^*, s_{-i}') = x_i(s_i^*, s_{-i}') p_i^* - \gamma_i z_i^* \geq \alpha_i$ implies

$$x_i(s_i^*, s_{-i}') \geq \frac{\alpha_i + \gamma_i z_i^*}{p_i^*} > \frac{\alpha_i}{C_i} > \frac{1}{2}, \tag{4.2}$$

where the last inequality is due to (4.1). If $z_i' = 0$ holds for a firm i, then $s_i' \notin C_i(s', \alpha_i)$ holds, since any strategy (z_i^*, p_i^*) achieving a profit of at least $\alpha_i > 0$ fulfills $z_i^* > 0$. Thus we can assume in the following that $z_i' > 0$ holds for all firms i. Then, since $n \geq 2$, there is at least one firm i with $x_i(s') \leq \frac{1}{2}$. We now show that $s_i' \notin C_i(s', \alpha_i)$ holds. If $z_i' = 0$ or $p_i' \leq \alpha_i$ holds, $s_i' \notin C_i(s', \alpha_i)$ follows, since $z_i^* > 0$ and $p_i^* > \alpha_i$ hold for any strategy (z_i^*, p_i^*) achieving a profit of at least α_i. Thus we can assume in the following that $z_i' > 0$ and $p_i' > \alpha_i$ hold. If $x_i(s') = 0$, the Wardrop equilibrium conditions yield $K(s') \leq b_i + p_i'$. Then, any strategy $\bar{s}_i = (\bar{z}_i, \bar{p}_i)$ with $\bar{p}_i \geq p_i'$ yields $x(\bar{s}_i, s_{-i}') = x(s')$, and thus $x_i(\bar{s}_i, s_{-i}') = 0$ and $\Pi_i(\bar{s}_i, s_{-i}') \leq 0 < \alpha_i$ hold. Therefore, $p_i^* < p_i'$ holds for any strategy (z_i^*, p_i^*)

achieving a profit of at least α_i, and $s_i' \notin C_i(s', \alpha_i)$ follows. We can thus assume in the following that $x_i(s') > 0$ holds. Summarizing, we can assume that the following inequalities are fulfilled:

$$0 < x_i(s_i', s_{-i}') \le \frac{1}{2}, \quad \alpha_i < p_i' \le C_i \quad \text{and} \quad 0 < z_i' < 1. \tag{4.3}$$

We now show that each strategy $s_i^* = (z_i^*, p_i^*)$ with $\Pi_i(s_i^*, s_{-i}') \ge \alpha_i$ fulfills $z_i^* > z_i'$, showing that $s_i' \notin C_i(s', \alpha_i)$ and completing the proof.

Assume, by contradiction, that there is a strategy $s_i^* = (z_i^*, p_i^*)$ which achieves a profit of at least α_i and fulfills $z_i^* \le z_i'$. For any strategy $\tilde{s}_i \in \mathcal{S}_i$, write $x(\tilde{s}_i) := x(\tilde{s}_i, s_{-i}')$ and $K(\tilde{s}_i) := K(\tilde{s}_i, s_{-i}')$. Now consider the strategy $\tilde{s}_i := (z_i', \alpha_i)$. Assume, for the moment, that

$$K(\tilde{s}_i) \le K(s_i^*) < K(s_i') \tag{4.4}$$

holds (we prove (4.4) below). Using $K(\tilde{s}_i) < K(s_i')$ then implies $\frac{a_i}{z_i'} x_i(\tilde{s}_i) + b_i + \alpha_i < \frac{a_i}{z_i'} x_i(s_i') + b_i + p_i'$. Reformulating this inequality and using (4.3) and (4.1) then yields

$$x_i(\tilde{s}_i) < \frac{z_i'(p_i' - \alpha_i)}{a_i} + x_i(s_i') < \frac{C_i - \alpha_i}{a_i} + \frac{1}{2} < \frac{\alpha_i}{C_i}. \tag{4.5}$$

The inequality $K(\tilde{s}_i) \le K(s_i^*)$ from (4.4) implies $x_j(\tilde{s}_i) \le x_j(s_i^*)$ for all firms $j \ne i$. Therefore, $x_i(\tilde{s}_i) \ge x_i(s_i^*)$ holds. Using $x_i(s_i^*) > \frac{\alpha_i}{C_i}$ from (4.2) now leads to $x_i(\tilde{s}_i) > \frac{\alpha_i}{C_i}$, which contradicts (4.5). To complete the proof, it remains to show (4.4). The property $K(s_i^*) < K(s_i')$ holds since $K(s_i^*) \ge K(s_i')$ would imply $x_j(s_i^*) \ge x_j(s_i')$ for all firms $j \ne i$, and thus $x_i(s_i^*) \le x_i(s_i')$, but we know from (4.2) and (4.3) that $x_i(s_i^*) > 1/2 \ge x_i(s_i')$. To prove the other inequality in (4.4), assume, by contradiction, that $K(\tilde{s}_i) > K(s_i^*)$. This implies $x_j(\tilde{s}_i) \ge x_j(s_i^*)$ for all firms $j \ne i$, and thus $x_i(\tilde{s}_i) \le x_i(s_i^*)$. Together with $z_i^* \le z_i'$ and $p_i^* > \alpha_i$, this leads to the following contradiction, and finally completes the proof:

$$K(\tilde{s}_i) \le \frac{a_i}{z_i'} x_i(\tilde{s}_i) + b_i + \alpha_i < \frac{a_i}{z_i^*} x_i(s_i^*) + b_i + p_i^* = K(s_i^*) < K(\tilde{s}_i).$$

$\square$

Using Theorem 4.4.3 together with Lemma 4.4.4 and Lemma 4.4.5 now yields the existence of a PNE:

Theorem 4.4.6 *Every capacity and price competition game has a PNE.*

Note here that in any PNE (z, p), there are at least two firms i with $z_i > 0$.

We conclude this section with an example showing that in general, a capacity and price competition game is not quasiconcave (that is, it is not the case that for all $i \in N$ and all $s_{-i} \in \mathcal{S}_{-i}$, the profit $\Pi_i(\cdot, s_{-i})$ is quasiconcave on $\mathcal{S}_i$). Thus, the result of Reny [95] cannot be used to show existence of PNE.

Example 4.4.7 Consider a capacity and price competition game with two firms, $N = \{1, 2\}$, and $a_1 = 1$, $a_2 = 2$, $b_1 = b_2 = 1$, $C_1 = C_2 = 10$, $\gamma_1 = \gamma_2 = \frac{1}{4}$. Assume that firm 1 chooses the strategy $s_1 = (z_1, p_1) = (1, 1)$, thus $\ell_1(x_1, z_1) + p_1 = x_1 + 2$.

We show that $\Pi_2(\cdot, s_1)$ is not quasiconcave on $\mathcal{S}_2$. To this end, note that any strategy (z_2, p_2) with $z_2 = 0$, $p_2 \in [0, 10]$ yields a profit of 0, thus in particular the strategy $(0, 10)$. Furthermore, if firm 2 chooses $(z_2, p_2) = (2, 1)$, this also results in a profit of $1/2 - 2/4 = 0$ (since $\ell_2(x_2, z_2) + p_2 = x_2 + 2$, and thus the induced Wardrop flow is $x_1 = x_2 = 1/2$). But the convex combination $\frac{1}{2}(0, 10) + \frac{1}{2}(2, 1) = (1, 11/2)$ yields profit $-1/4 < 0$ (since $x_1 = 1$, $x_2 = 0$ is the induced Wardrop flow). By definition, this shows that $\Pi_2(\cdot, s_1)$ is not quasiconcave on $\mathcal{S}_2$.

4.5 Uniqueness of Equilibria

As we have seen in the last section, a capacity and price competition game always has a PNE. In this section we show that this equilibrium is *essentially* unique. With *essentially* we mean that if (z, p) and (z', p') are two different PNE, and $i \in N$ is a firm such that $(z_i, p_i) \neq (z_i', p_i')$, then $z_i = z_i' = 0$ holds (and thus $p_i \neq p_i'$).

For a PNE $s = (z, p)$, denote by $N^+(z, p) := \{i \in N : z_i > 0\}$ the set of firms with positive capacity (note that $|N^+(z, p)| \geq 2$ and $N^+(z, p) = \{i \in N : x_i(s) > 0\}$). For $i \in N^+(z, p)$, let $(P_i^1)(s_{-i})$ and $(P_i^2)(s_{-i})$ be the two auxiliary problems from Section 4.3.[4] By Lemma 4.3.4, the routing cost $K(z, p)$ is an optimal solution of

---

[4]In Section 4.3, we considered *fixed* strategies s_{-i}, thus we just used (P_i^1) and (P_i^2) for the problems corresponding to s_{-i}. In this section, we need to consider different strategy profiles, thus we now write $(P_i^1)(s_{-i})$ and $(P_i^2)(s_{-i})$, as well as $K_i^{\max}(s_{-i})$.

either $(P_i^1)(s_{-i})$ or $(P_i^2)(s_{-i})$. We denote by $N_1^+(z, p)$ the set of firms $i \in N^+(z, p)$ such that $K(z, p)$ is an optimal solution of $(P_i^1)(s_{-i})$, and $N_2^+(z, p)$ contains the firms $i \in N^+(z, p)$ such that $K(z, p)$ is an optimal solution of $(P_i^2)(s_{-i})$. Thus $N^+(z, p) = N_1^+(z, p) \,\dot\cup\, N_2^+(z, p)$. Throughout this section, we use the simplified notation $N' \setminus i$ instead of $N' \setminus \{i\}$ for any subset $N' \subseteq N$ of firms and $i \in N'$.

Note that the proofs in this section are similar to the proofs that Johari et al. [75] use to derive their uniqueness results. However, since our model includes price caps, some new ideas are required, in particular the decomposition of $N^+(z, p)$ in $N_1^+(z, p) \,\dot\cup\, N_2^+(z, p)$.

We first derive further necessary equilibrium conditions (by using the KKT conditions) which will become useful in the following analysis.

Lemma 4.5.1 *Let $s = (z, p)$ be a PNE with $x := x(z, p)$ and $K := K(z, p)$. Let $i \in N^+ := N^+(z, p)$. If $p_i < C_i$ holds, then $z_i = \sqrt{\frac{a_i}{\gamma_i}} x_i$, $p_i = \frac{x_i}{\sum_{j \in N^+ \setminus i} \frac{z_j}{a_j}} + \sqrt{a_i \gamma_i}$, and if $p_i = C_i$, then $z_i = \frac{a_i x_i}{K - b_i - C_i}$ and $\frac{C_i}{1 + \frac{a_i}{z_i} \cdot \sum_{j \in N^+ \setminus i} \frac{z_j}{a_j}} = \frac{\gamma_i z_i^2}{a_i x_i \sum_{j \in N^+ \setminus i} \frac{z_j}{a_j}}$.*

Proof Since (z, p) is a PNE, $p_j > 0$ and $x_j > 0$ holds for all $j \in N^+$, and $x_k = 0$ holds for $k \notin N^+$. Furthermore, (z_i, p_i) is a best response of firm i to s_{-i} and $K = \frac{a_j x_j}{z_j} + b_j + p_j$ holds for all $j \in N^+$. Altogether we get that $(z_i, p_i, (x_j)_{j \in N^+})$ is an optimal solution for the following optimization problem (with variables $(z_i', p_i', (x_j')_{j \in N^+})$):

$$\max \quad x_i' p_i' - \gamma_i z_i'$$
$$\text{subject to} \quad 0 \le p_i' \le C_i$$
$$0 < z_i'$$
$$\sum_{j \in N^+} x_j' = 1$$
$$x_j' \ge 0 \,\forall j \in N^+$$
$$\frac{a_i x_i'}{z_i'} + b_i + p_i' = \frac{a_j x_j'}{z_j} + b_j + p_j \,\forall j \in N^+ \setminus i.$$

It is easy to show that the LICQ is fulfilled for $(z_i, p_i, (x_j)_{j \in N^+}) = (z_i, p_i, x_i, (x_j)_{j \in N^+ \setminus i})$ (see the end of the proof), thus the KKT conditions are fulfilled. We get the following equations:

$$\gamma_i - \frac{a_i x_i}{z_i^2} \sum_{j \in N^+ \setminus i} \lambda_j = 0 \tag{4.6}$$

$$-x_i + \mu + \sum_{j \in N^+ \setminus i} \lambda_j = 0 \tag{4.7}$$

$$-p_i + \lambda + \frac{a_i}{z_i} \sum_{j \in N^+ \setminus i} \lambda_j = 0 \tag{4.8}$$

$$\lambda - \lambda_j \frac{a_j}{z_j} = 0 \; \forall j \in N^+ \setminus i. \tag{4.9}$$

We now distinguish between the two cases $p_i < C_i$ and $p_i = C_i$.

In the first case, $\mu = 0$ holds, and (4.7) yields $x_i = \sum_{j \in N^+ \setminus i} \lambda_j$. Using this, (4.6) yields $z_i = \sqrt{\frac{a_i}{\gamma_i}} x_i$. Plugging this in (4.8) leads to $p_i = \lambda + \frac{a_i x_i}{z_i} = \lambda + \sqrt{a_i \gamma_i}$. Using (4.9), i.e., $\lambda_j = \lambda \frac{z_j}{a_j}$ for all $j \in N^+ \setminus i$, together with (4.7) yields $x_i = \lambda \sum_{j \in N^+ \setminus i} \frac{z_j}{a_j}$, or equivalently, $\lambda = \frac{x_i}{\sum_{j \in N^+ \setminus i} \frac{z_j}{a_j}}$. This shows $p_i = \frac{x_i}{\sum_{j \in N^+ \setminus i} \frac{z_j}{a_j}} + \sqrt{a_i \gamma_i}$, as required.

The other case is $p_i = C_i$. The formula for z_i follows from $K = \frac{a_i x_i}{z_i} + b_i + C_i$. Plugging $\lambda_j = \lambda \frac{z_j}{a_j}$ for all $j \in N^+ \setminus i$ in (4.6) and (4.8) yields

$$\lambda = \frac{\gamma_i z_i^2}{a_i x_i \sum_{j \in N^+ \setminus i} \frac{z_j}{a_j}}$$

and

$$\lambda = C_i - \frac{a_i}{z_i} \lambda \cdot \sum_{j \in N^+ \setminus i} \frac{z_j}{a_j} \Leftrightarrow \lambda = \frac{C_i}{1 + \frac{a_i}{z_i} \sum_{j \in N^+ \setminus i} \frac{z_j}{a_j}},$$

which shows the desired equality.

It remains to show that the LICQ is fulfilled for $(z_i, p_i, (x_j)_{j \in N^+}) = (z_i, p_i, x_i, (x_j)_{j \in N^+ \setminus i})$. Besides the gradients of the equalities, we have to take into account the gradient of the inequality $p_i' \leq C_i$ (all other inequalities are not tight). Thus consider

$$
\alpha_1 \cdot \begin{pmatrix} 0 \\ 1 \\ 0 \\ 0 \\ \vdots \\ \vdots \\ \vdots \\ \vdots \\ 0 \end{pmatrix} + \alpha_2 \cdot \begin{pmatrix} 0 \\ 0 \\ 1 \\ 1 \\ \vdots \\ \vdots \\ \vdots \\ \vdots \\ 1 \end{pmatrix} + \sum_{j \in N^+ \setminus i} \alpha_j \cdot \begin{pmatrix} -\frac{a_i x_i}{z_i^2} \\ 1 \\ \frac{a_i}{z_i} \\ 0 \\ \vdots \\ 0 \\ -\frac{a_j}{z_j} \\ 0 \\ \vdots \\ 0 \end{pmatrix} = 0,
$$

where all vectors are in $\mathbb{R}^{2+|N^+|}$. For $j \in N^+ \setminus i$ and the corresponding vector in the sum, the entry corresponding to j is $-a_j/z_j$, and the entries corresponding to $N^+ \setminus \{i, j\}$ are all zero. We show that all α_k are zero. Considering the first row yields $-\frac{a_i x_i}{z_i^2} \cdot \sum_{j \in N^+ \setminus i} \alpha_j = 0$ which yields $\sum_{j \in N^+ \setminus i} \alpha_j = 0$ since $-\frac{a_i x_i}{z_i^2} < 0$. Using this, we get from the second row $\alpha_1 + \sum_{j \in N^+ \setminus i} \alpha_j = \alpha_1 = 0$. The third row yields $\alpha_2 + \frac{a_i}{z_i} \cdot \sum_{j \in N^+ \setminus i} \alpha_j = \alpha_2 = 0$. Finally, any row corresponding to $j \in N^+ \setminus \{i\}$ reads $\alpha_2 + \alpha_j \cdot (-\frac{a_j}{z_j}) = \alpha_j \cdot (-\frac{a_j}{z_j}) = 0$ and this shows $\alpha_j = 0$ since $\frac{a_j}{z_j} > 0$. $\square$

In the next lemma, we introduce two functions Γ_i^1 and Γ_i^2 for each firm i and derive useful properties of these functions.

Lemma 4.5.2 *For each $i \in N$, define*

$$
\Gamma_i^1 : (\sqrt{a_i \gamma_i} + b_i, \infty) \to \mathbb{R}, \quad \Gamma_i^1(\kappa) := \frac{\sqrt{a_i \gamma_i}}{\kappa - \sqrt{a_i \gamma_i} - b_i}
$$

and

$$
\Gamma_i^2 : (b_i + C_i, \infty) \to \mathbb{R}, \quad \Gamma_i^2(\kappa) := \frac{a_i \gamma_i / C_i}{\kappa - b_i - C_i}.
$$

Furthermore, let $s := (z, p)$ be a PNE with $x := x(z, p)$, cost $K := K(z, p)$, and $N^+ := N^+(z, p)$ with $N_1^+ := N_1^+(z, p)$, $N_2^+ := N_2^+(z, p)$. Then:

1. Γ_i^1 and Γ_i^2 are strictly decreasing functions.

2. *If* $i \in N_1^+$, *then* $\Gamma_i^1(K) = 1 - \dfrac{\frac{z_i}{a_i}}{\sum_{j \in N^+} \frac{z_j}{a_j}} < 1.$

3. *If* $i \in N_2^+$, *then* $\Gamma_i^2(K) = 1 - \dfrac{\frac{z_i}{a_i}}{\sum_{j \in N^+} \frac{z_j}{a_j}} < 1.$

4. $\sum_{i \in N_1^+} \Gamma_i^1(K) + \sum_{i \in N_2^+} \Gamma_i^2(K) = |N^+| - 1.$

5. *If* $i \in N_1^+$ *and there is a (different) PNE* $s' = (z', p')$ *with* $i \in N_2^+(z', p')$, *then* $K < K' := K(z', p')$ *and* $\Gamma_i^1(K) > \Gamma_i^2(K').$

Proof Statement 1. is clear from the definitions of Γ_i^1 and Γ_i^2, so turn to statement 2. and let $i \in N_1^+$. Lemma 4.3.4 yields $z_i = \sqrt{\frac{a_i}{\gamma_i}} x_i$, $p_i = K - \sqrt{a_i \gamma_i} - b_i$. Using Lemma 4.5.1, we get $K - \sqrt{a_i \gamma_i} - b_i = \dfrac{x_i}{\sum_{j \in N^+ \setminus i} \frac{z_j}{a_j}} + \sqrt{a_i \gamma_i}$, which is equivalent to

$$K = \frac{x_i}{\sum_{j \in N^+ \setminus i} \frac{z_j}{a_j}} + 2\sqrt{a_i \gamma_i} + b_i.$$

(Note that $p_i = C_i$ is possible (namely if $K = \sqrt{a_i \gamma_i} + b_i + C_i$), but Lemma 4.5.1 yields the stated equality also for this case.) Now define

$$B := \sum_{j \in N^+} \frac{z_j}{a_j}.$$

Using $z_i = \sqrt{\frac{a_i}{\gamma_i}} x_i$, we rewrite K as follows:

$$
\begin{aligned}
K &= \frac{x_i}{B - \frac{z_i}{a_i}} + 2\sqrt{a_i \gamma_i} + b_i \\
&= \frac{x_i}{B - \frac{x_i}{\sqrt{a_i \gamma_i}}} + 2\sqrt{a_i \gamma_i} + b_i \\
&= \frac{x_i + (B - \frac{x_i}{\sqrt{a_i \gamma_i}})\sqrt{a_i \gamma_i}}{B - \frac{x_i}{\sqrt{a_i \gamma_i}}} + \sqrt{a_i \gamma_i} + b_i \\
&= \frac{B\sqrt{a_i \gamma_i}}{B - \frac{x_i}{\sqrt{a_i \gamma_i}}} + \sqrt{a_i \gamma_i} + b_i
\end{aligned}
$$

Using this we get statement 2.:

$$\Gamma_i^1(K) = \frac{B - \frac{x_i}{\sqrt{a_i \gamma_i}}}{B} = 1 - \frac{\frac{z_i}{a_i}}{B} < 1.$$

For statement 3., let $i \in N_2^+$. Lemma 4.3.4 and Lemma 4.5.1 imply

$$\frac{C_i}{1 + \frac{a_i}{z_i} \cdot \sum_{j \in N^+ \setminus i} \frac{z_j}{a_j}} = \frac{\gamma_i z_i^2}{a_i x_i \sum_{j \in N^+ \setminus i} \frac{z_j}{a_j}}.$$

Rearranging and using the definition of B yields

$$\frac{a_i x_i}{z_i} = \frac{\gamma_i z_i (1 + \frac{a_i}{z_i} \cdot \sum_{j \in N^+ \setminus i} \frac{z_j}{a_j})}{C_i \cdot \sum_{j \in N^+ \setminus i} \frac{z_j}{a_j}} = \frac{\gamma_i z_i (1 + \frac{a_i}{z_i} \cdot (B - \frac{z_i}{a_i}))}{C_i \cdot (B - \frac{z_i}{a_i})} = \frac{\gamma_i a_i B}{C_i \cdot (B - \frac{z_i}{a_i})}.$$

Using $K = \frac{a_i x_i}{z_i} + b_i + C_i$ then yields

$$K = \frac{\gamma_i a_i B}{C_i \cdot (B - \frac{z_i}{a_i})} + b_i + C_i$$

and thus statement 3. follows:

$$\Gamma_i^2(K) = \frac{B - \frac{z_i}{a_i}}{B} = 1 - \frac{\frac{z_i}{a_i}}{B} < 1.$$

Statement 4. now follows from the statements 2. and 3.:

$$\sum_{i \in N_1^+} \Gamma_i^1(K) + \sum_{i \in N_2^+} \Gamma_i^2(K) = \sum_{i \in N_1^+} \left(1 - \frac{\frac{z_i}{a_i}}{B}\right) + \sum_{i \in N_2^+} \left(1 - \frac{\frac{z_i}{a_i}}{B}\right)$$

$$= |N^+| - \frac{1}{B} \cdot \sum_{j \in N^+} \frac{z_j}{a_j} = |N^+| - 1.$$

It remains to show statement 5. Let $i \in N_1^+ \cap N_2^+(z', p')$. Since $i \in N_1^+$, the cost K is in particular feasible for $(P_i^1)(s_{-i})$, thus $2\sqrt{a_i \gamma_i} + b_i \le K \le \sqrt{a_i \gamma_i} + b_i + C_i$ holds. Analogously, using $i \in N_2^+(z', p')$, the cost K' is feasible for $(P_i^2)(s_{-i}')$, therefore $\sqrt{a_i \gamma_i} + b_i + C_i < K' < K_i^{\max}(s_{-i}')$. Together we get $K \le \sqrt{a_i \gamma_i} + b_i + C_i < K'$. It remains to show $\Gamma_i^1(K) > \Gamma_i^2(K')$. By definition of $N_2^+(z', p')$, the cost K'

is an optimal solution for problem $(P_i^2)(s'_{-i})$ and in particular (see 1. and 3. of Lemma 4.3.5) yields a better objective function value than $\sqrt{a_i \gamma_i} + b_i + C_i$ in $(P_i^1)(s'_{-i})$ (note that $\sqrt{a_i \gamma_i} + b_i + C_i$ is feasible for $(P_i^1)(s'_{-i})$ since $2\sqrt{a_i \gamma_i} + b_i \leq K \leq \sqrt{a_i \gamma_i} + b_i + C_i < K' < K_i^{\max}(s'_{-i}))$. If $\overline{x}_i$ denotes the function occurring in the definitions of $(P_i^1)(s'_{-i})$ and $(P_i^2)(s'_{-i})$, we thus get

$$\overline{x}_i(\sqrt{a_i \gamma_i} + b_i + C_i) \cdot (C_i - \sqrt{a_i \gamma_i}) < \overline{x}_i(K') \cdot (C_i - \frac{a_i \gamma_i}{K' - b_i - C_i})$$
$$\leq \overline{x}_i(\sqrt{a_i \gamma_i} + b_i + C_i) \cdot (C_i - \frac{a_i \gamma_i}{K' - b_i - C_i}),$$

where the last inequality follows from $\sqrt{a_i \gamma_i} + b_i + C_i < K'$ and the fact that $\overline{x}_i$ is a decreasing function. Since $\overline{x}_i(\sqrt{a_i \gamma_i} + b_i + C_i) > 0$, we get

$$C_i - \sqrt{a_i \gamma_i} < C_i - \frac{a_i \gamma_i}{K' - b_i - C_i},$$

or equivalently

$$\sqrt{a_i \gamma_i} > \frac{a_i \gamma_i}{K' - b_i - C_i},$$

which yields

$$\frac{\sqrt{a_i \gamma_i}}{C_i} > \frac{\frac{a_i \gamma_i}{C_i}}{K' - b_i - C_i} = \Gamma_i^2(K').$$

Note that $2\sqrt{a_i \gamma_i} + b_i \leq K \leq \sqrt{a_i \gamma_i} + b_i + C_i$ and thus

$$\frac{1}{K - \sqrt{a_i \gamma_i} - b_i} \geq \frac{1}{C_i},$$

which leads to

$$\Gamma_i^1(K) = \frac{\sqrt{a_i \gamma_i}}{K - \sqrt{a_i \gamma_i} - b_i} \geq \frac{\sqrt{a_i \gamma_i}}{C_i} > \Gamma_i^2(K'),$$

as desired. $\square$

Now we turn to the desired uniqueness of the equilibrium. We start with the following lemma.

Lemma 4.5.3 *For a fixed subset N^+ of the firms and a fixed disjoint decomposition $N^+ = N_1^+ \cup N_2^+$, there is essentially at most one PNE (z, p) such that $N^+(z, p) = N^+$, $N_1^+(z, p) = N_1^+$ and $N_2^+(z, p) = N_2^+$.*

Proof Assume that there are two PNE (z, p) and (z', p') with the described properties, i.e., $N^+(z, p) = N^+(z', p') = N^+$, $N_1^+(z, p) = N_1^+(z', p') = N_1^+$ and $N_2^+(z, p) = N_2^+(z', p') = N_2^+$. Let $x := x(z, p)$ and $x' := x(z', p')$ with costs $K := K(z, p)$ and $K' := K(z', p')$. We show that $(z_i, p_i) = (z_i', p_i')$ holds for all $i \in N^+$, showing that (z, p) and (z', p') are essentially the same.

First note that $K = K'$ holds, since $f(\kappa) := \sum_{i \in N_1^+} \Gamma_i^1(\kappa) + \sum_{i \in N_2^+} \Gamma_i^2(\kappa)$ is a strictly decreasing function in κ and

$$f(K) = \sum_{i \in N_1^+} \Gamma_i^1(K) + \sum_{i \in N_2^+} \Gamma_i^2(K) = |N^+| - 1 = \sum_{i \in N_1^+} \Gamma_i^1(K') + \sum_{i \in N_2^+} \Gamma_i^2(K') = f(K')$$

holds from 4. in Lemma 4.5.2. This implies $p_i = p_i'$ for all $i \in N^+$, since $p_i = K - \sqrt{a_i \gamma_i} - b_i = K' - \sqrt{a_i \gamma_i} - b_i = p_i'$ holds for $i \in N_1^+$, and $p_i = C_i = p_i'$ for $i \in N_2^+$.

If $B := \sum_{j \in N^+} \frac{z_j}{a_j} = \sum_{j \in N^+} \frac{z_j'}{a_j} =: B'$ holds, we also get $z_i = z_i'$ for all $i \in N^+$, since 2. of Lemma 4.5.2 yields

$$z_i = (1 - \Gamma_i^1(K)) a_i B = (1 - \Gamma_i^1(K')) a_i B' = z_i'$$

for all $i \in N_1^+$, and 3. of Lemma 4.5.2 yields

$$z_i = (1 - \Gamma_i^2(K)) a_i B = (1 - \Gamma_i^2(K')) a_i B' = z_i'$$

for all $i \in N_2^+$.

It remains to show $B = B'$. First consider $i \in N_1^+$. Using $z_i = \sqrt{a_i / \gamma_i} \cdot x_i$ and $z_i' = \sqrt{a_i / \gamma_i} \cdot x_i'$, as well as 2. of Lemma 4.5.2, yields

$$x_i / B = (1 - \Gamma_i^1(K)) \sqrt{a_i \gamma_i} = (1 - \Gamma_i^1(K')) \sqrt{a_i \gamma_i} = x_i' / B'.$$

For $i \in N_2^+$, we use $\frac{z_i}{a_i} = \frac{x_i}{K - b_i - C_i}$ and $\frac{z_i'}{a_i} = \frac{x_i'}{K' - b_i - C_i}$ and 3. of Lemma 4.5.2 to achieve

$$x_i / B = (1 - \Gamma_i^2(K))(K - b_i - C_i) = (1 - \Gamma_i^2(K'))(K' - b_i - C_i) = x_i' / B'.$$

Altogether we have $x_i/B = x_i'/B'$ for all $i \in N^+$. Using that $\sum_{i \in N^+} x_i = 1 = \sum_{i \in N^+} x_i'$ yields $B = B'$, as desired:

$$\frac{1}{B} = \sum_{i \in N^+} \frac{x_i}{B} = \sum_{i \in N^+} \frac{x_i'}{B'} = \frac{1}{B'}.$$

$\square$

In the previous lemma, we showed that given a fixed subset $N^+ \subseteq N$ and a fixed disjoint decomposition $N^+ = N_1^+ \,\dot\cup\, N_2^+$, there is at most one PNE (z, p) such that $N^+(z, p) = N^+$, $N_1^+(z, p) = N_1^+$ and $N_2^+(z, p) = N_2^+$. Next, we strengthen this result by showing that for a fixed subset $N^+ \subseteq N$, there is at most one PNE (z, p) with $N^+(z, p) = N^+$ (independently of the decomposition of N^+).

Lemma 4.5.4 *For a fixed subset N^+ of the firms, there is essentially at most one PNE (z, p) with $N^+(z, p) = N^+$.*

Proof Assume, by contradiction, that there are two essentially different PNE (z, p) and $(\overline{z}, \overline{p})$ with $N^+(z, p) = N^+ = N^+(\overline{z}, \overline{p})$. Let $N_1^+ := N_1^+(z, p)$, $N_2^+ := N_2^+(z, p)$ and $\overline{N}_1^+ := N_1^+(\overline{z}, \overline{p})$, $\overline{N}_2^+ := N_2^+(\overline{z}, \overline{p})$. Further denote $x := x(z, p)$, $K := K(z, p)$ and $\overline{x} := x(\overline{z}, \overline{p})$, $\overline{K} := K(\overline{z}, \overline{p})$.

Lemma 4.5.3 yields that the decompositions of N^+ have to be different. Without loss of generality, there is a firm $j \in N_1^+ \setminus \overline{N}_1^+$. Since $j \in \overline{N}_2^+$, statement 5. of Lemma 4.5.2 yields $K < \overline{K}$. The existence of a firm $i \in \overline{N}_1^+ \setminus N_1^+$ leads (by the same argumentation) to the contradiction $\overline{K} < K$, thus $\overline{N}_1^+ \subsetneq N_1^+$ and $N_2^+ \subsetneq \overline{N}_2^+$ hold and we can write (using 4. of Lemma 4.5.2)

$$|N^+| - 1 = \sum_{i \in N_1^+} \Gamma_i^1(K) + \sum_{i \in N_2^+} \Gamma_i^2(K) = \sum_{i \in N_1^+ \setminus \overline{N}_1^+} \Gamma_i^1(K) + \sum_{i \in \overline{N}_1^+} \Gamma_i^1(K) + \sum_{i \in N_2^+} \Gamma_i^2(K)$$

and

$$|N^+| - 1 = \sum_{i \in \overline{N}_1^+} \Gamma_i^1(\overline{K}) + \sum_{i \in \overline{N}_2^+} \Gamma_i^2(\overline{K}) = \sum_{i \in \overline{N}_1^+} \Gamma_i^1(\overline{K}) + \sum_{i \in \overline{N}_2^+ \setminus N_2^+} \Gamma_i^2(\overline{K}) + \sum_{i \in N_2^+} \Gamma_i^2(\overline{K}).$$

Using that $K < \overline{K}$ and $\sum_{i \in N_2^+} \Gamma_i^2(\kappa)$ is a decreasing function in κ yields

$$\sum_{i \in N_2^+} \Gamma_i^2(K) \geq \sum_{i \in N_2^+} \Gamma_i^2(\overline{K}).$$

Furthermore, $\sum_{i \in \overline{N}_1^+} \Gamma_i^1(\kappa)$ is also decreasing in κ, thus

$$\sum_{i \in \overline{N}_1^+} \Gamma_i^1(K) \geq \sum_{i \in \overline{N}_1^+} \Gamma_i^1(\overline{K}).$$

Finally, statement 5. of Lemma 4.5.2 yields $\Gamma_i^1(K) > \Gamma_i^2(\overline{K})$ for all $i \in N_1^+ \setminus \overline{N}_1^+ = \overline{N}_2^+ \setminus N_2^+ \neq \emptyset$, thus

$$\sum_{i \in N_1^+ \setminus \overline{N}_1^+} \Gamma_i^1(K) > \sum_{i \in \overline{N}_2^+ \setminus N_2^+} \Gamma_i^2(\overline{K})$$

holds. Altogether we get the contradiction

$$|N^+| - 1 = \sum_{i \in N_1^+ \setminus \overline{N}_1^+} \Gamma_i^1(K) + \sum_{i \in \overline{N}_1^+} \Gamma_i^1(K) + \sum_{i \in N_2^+} \Gamma_i^2(K)$$

$$> \sum_{i \in \overline{N}_1^+} \Gamma_i^1(\overline{K}) + \sum_{i \in \overline{N}_2^+ \setminus N_2^+} \Gamma_i^2(\overline{K}) + \sum_{i \in N_2^+} \Gamma_i^2(\overline{K})$$

$$= |N^+| - 1,$$

which completes the proof. $\qquad\square$

For the desired uniqueness of the PNE, it remains to show that there is at most one set N^+ such that a PNE (z, p) with $N^+(z, p) = N^+$ exists. To this end, we first show that each firm i has a certain threshold K_i^* such that, for any PNE (z, p), firm i has $z_i > 0$ if and only if $K(z, p) > K_i^*$.

Lemma 4.5.5 *For each $i \in N$, define*

$$K_i^* := \begin{cases} \frac{a_i \gamma_i}{C_i} + b_i + C_i, & \text{if } \sqrt{a_i \gamma_i} > C_i, \\ 2\sqrt{a_i \gamma_i} + b_i, & \text{else.} \end{cases}$$

Then for any PNE $s = (z, p)$ and any firm $i \in N$, it holds that $z_i > 0$ if and only if $K(z, p) > K_i^*$.

Proof Let $s = (z, p)$ be a PNE with $x := x(z, p)$, $K := K(z, p)$, and $i \in N$.

First assume that $z_i = 0$. Since (z, p) is a PNE, the strategy $(z_i, p_i) = (0, p_i)$ is a best response of firm i to s_{-i}. As we have seen in Theorem 4.3.6, this is equivalent to the fact that both problems $(P_i^1)(s_{-i})$ and $(P_i^2)(s_{-i})$ are infeasible. Note that $K = K_i^{\max}(s_{-i})$ holds due to the definition of $K_i^{\max}(s_{-i})$ (cf. page 192) and

$$0 = 1 - \sum_{j \in N : x_j > 0} x_j = 1 - \sum_{j \in N \setminus i : z_j > 0, b_j + p_j < K} \frac{(K - b_j - p_j)z_j}{a_j},$$

where we used that $\{j \in N : x_j > 0\} = \{j \in N \setminus i : z_j > 0, b_j + p_j < K\}$. To show $K \le K_i^*$, we have to distinguish between the two cases $\sqrt{a_i \gamma_i} > C_i$ and $\sqrt{a_i \gamma_i} \le C_i$. First consider $\sqrt{a_i \gamma_i} > C_i$, thus $K_i^* = a_i \gamma_i / C_i + b_i + C_i$. Since $(P_i^2)(s_{-i})$ is infeasible and $\sqrt{a_i \gamma_i} + b_i + C_i < a_i \gamma_i / C_i + b_i + C_i$, we get the desired inequality $K = K_i^{\max}(s_{-i}) \le a_i \gamma_i / C_i + b_i + C_i = K_i^*$. Now consider $\sqrt{a_i \gamma_i} \le C_i$, i.e., $K_i^* = 2\sqrt{a_i \gamma_i} + b_i$. Since $(P_i^1)(s_{-i})$ is infeasible and $2\sqrt{a_i \gamma_i} + b_i \le \sqrt{a_i \gamma_i} + b_i + C_i$, we get $K = K_i^{\max}(s_{-i}) \le 2\sqrt{a_i \gamma_i} + b_i = K_i^*$, as desired. We have seen that $z_i = 0$ implies $K \le K_i^*$, or, equivalently, $K > K_i^*$ implies $z_i > 0$.

It remains to show the other direction, i.e., $z_i > 0$ implies $K > K_i^*$. We consider the two cases $\sqrt{a_i \gamma_i} > C_i$ and $\sqrt{a_i \gamma_i} \le C_i$ and use our results from Lemma 4.3.4 and Theorem 4.3.6. If $\sqrt{a_i \gamma_i} > C_i$, thus $K_i^* = \frac{a_i \gamma_i}{C_i} + b_i + C_i$, the cost K is an optimal solution for problem $(P_i^2)(s_{-i})$ with positive objective function value (note that $(P_i^1)(s_{-i})$ is infeasible), therefore $K_i^* = \frac{a_i \gamma_i}{C_i} + b_i + C_i < K$. In the second case, i.e., $\sqrt{a_i \gamma_i} \le C_i$ and $K_i^* = 2\sqrt{a_i \gamma_i} + b_i$, the cost K either is optimal for $(P_i^1)(s_{-i})$, or optimal for $(P_i^2)(s_{-i})$, and has positive objective function value in both cases. We get the desired property, since $K_i^* = 2\sqrt{a_i \gamma_i} + b_i < K$ holds for the first case, and $K_i^* = 2\sqrt{a_i \gamma_i} + b_i \le \sqrt{a_i \gamma_i} + b_i + C_i < K$ holds for the second case, completing the proof. $\square$

We can now show the remaining result for the desired uniqueness of PNE.

Lemma 4.5.6 *There is at most one subset N^+ of the firms such that a PNE (z, p) with $N^+(z, p) = N^+$ exists.*

Proof Assume, by contradiction, that there are two different subsets N^+ und $\overline{N}^+$ with corresponding PNE (z, p) and $(\overline{z}, \overline{p})$, such that $N^+(z, p) = N^+$ and $N^+(\overline{z}, \overline{p}) = \overline{N}^+$. Let $N_1^+ \dot\cup N_2^+$ and $\overline{N}_1^+ \dot\cup \overline{N}_2^+$ be the decompositions of N^+ and $\overline{N}^+$,

that is, $N_1^+(z, p) = N_1^+$, $N_2^+(z, p) = N_2^+$, $N_1^+(\overline{z}, \overline{p}) = \overline{N}_1^+$ and $N_2^+(\overline{z}, \overline{p}) = \overline{N}_2^+$.
Finally, denote $x := x(z, p)$, $K := K(z, p)$ and $\overline{x} := x(\overline{z}, \overline{p})$, $\overline{K} := K(\overline{z}, \overline{p})$.

Using Lemma 4.5.5, we can assume w.l.o.g. that $K < \overline{K}$ and $N^+ \subsetneq \overline{N}^+$. Then, $N_2^+ \subseteq \overline{N}_2^+$ holds, since the existence of a firm $i \in N_2^+ \setminus \overline{N}_2^+$, i.e., $i \in N_2^+ \cap \overline{N}_1^+$, leads to the contradiction $\overline{K} < K$ by statement 5. of Lemma 4.5.2. Furthermore, if there is a firm $i \in N_1^+ \setminus \overline{N}_1^+$, i.e., $i \in N_1^+ \cap \overline{N}_2^+$, statement 5. of Lemma 4.5.2 yields $\Gamma_i^1(K) > \Gamma_i^2(\overline{K})$. Finally, $\Gamma_i^1(\overline{K}) < 1$ holds for all $i \in \overline{N}_1^+$, and $\Gamma_i^2(\overline{K}) < 1$ holds for all $i \in \overline{N}_2^+$ (see 2. and 3. of Lemma 4.5.2). Altogether, this leads to the following contradiction, and completes the proof (where we additionally use $K < \overline{K}$, and the statements 1. and 4. of Lemma 4.5.2):

$$
\begin{aligned}
|\overline{N}^+| - 1 &= \sum_{i \in \overline{N}_1^+} \Gamma_i^1(\overline{K}) + \sum_{i \in \overline{N}_2^+} \Gamma_i^2(\overline{K}) \\
&= \sum_{i \in \overline{N}_1^+ \cap N_1^+} \Gamma_i^1(\overline{K}) + \sum_{i \in \overline{N}_1^+ \setminus N^+} \Gamma_i^1(\overline{K}) + \sum_{i \in \overline{N}_2^+ \cap N_1^+} \Gamma_i^2(\overline{K}) \\
&\quad + \sum_{i \in N_2^+} \Gamma_i^2(\overline{K}) + \sum_{i \in \overline{N}_2^+ \setminus N^+} \Gamma_i^2(\overline{K}) \\
&< \sum_{i \in N_1^+ \cap \overline{N}_1^+} \Gamma_i^1(K) + \sum_{i \in N_1^+ \cap \overline{N}_2^+} \Gamma_i^1(K) + \sum_{i \in N_2^+} \Gamma_i^2(K) + |\overline{N}^+| - |N^+| \\
&= \sum_{i \in N_1^+} \Gamma_i^1(K) + \sum_{i \in N_2^+} \Gamma_i^2(K) + |\overline{N}^+| - |N^+| \\
&= |N^+| - 1 + |\overline{N}^+| - |N^+| = |\overline{N}^+| - 1.
\end{aligned}
$$

$\square$

Together with the existence result in Theorem 4.4.6, the preceding Lemma 4.5.4 and Lemma 4.5.6 show that there is essentially one PNE.

Theorem 4.5.7 *Every capacity and price competition game has an (essentially) unique PNE, i.e., if (z, p) and (z', p') are two different PNE and $i \in N$ is a firm such that $(z_i, p_i) \neq (z_i', p_i')$, then $z_i = z_i' = 0$ holds.*

4.6 Quality of Equilibria

In the last section, we showed that a capacity and price competition game has an (essentially) unique PNE. Now we show that this PNE can be arbitrarily inefficient compared to a *social optimum*.

Define the *social cost* $C(z, p)$ of a strategy profile $s = (z, p)$ as

$$C(z, p) = \begin{cases} \sum_{i \in \{1,\dots,n\}: z_i > 0} (\ell_i(x_i(s), z_i) x_i(s) + \gamma_i z_i), & \text{if } \sum_{i=1}^{n} z_i > 0, \\ \infty, & \text{else.} \end{cases}$$

The function $C(z, p)$ measures utilitarian social welfare over firms and customers (the price component cancels out).

Since a capacity and price competition game G has an essentially unique PNE, all PNE of G have the same social cost. If we denote this cost by $C(\text{PNE}(G))$, we get

$$\text{PoA}(G) = \text{PoS}(G) = \frac{C(\text{PNE}(G))}{\text{OPT}(G)},$$

where $\text{OPT}(G) := \min\{C(s) : s \in \mathcal{S}\}$ denotes the minimum social cost in G (compared to all possible strategy profiles).

The following theorem shows that the worst-case PoA and PoS are unbounded for capacity and price competition games:

Theorem 4.6.1 *Let $\mathcal{G}$ be the class of capacity and price competition games. Then* $\text{PoA}(\mathcal{G}) = \text{PoS}(\mathcal{G}) = \infty$. *This even holds for games with only two firms.*

Proof Consider the capacity and price competition game G_M with

$$n = 2 \text{ and } a_1 = \gamma_1 = C_1 = 1, b_1 = 0 \text{ and } a_2 = \gamma_2 = C_2 = M, b_2 = 0,$$

where $M \geq 1$. By $z_1 = 1, p_1 = z_2 = p_2 = 0$, we get a profile with social cost 2, thus $\text{OPT}(G_M) \leq 2$. We will now show that $C(\text{PNE}(G_M)) > M$ holds, which implies

$$\text{PoA}(\mathcal{G}) = \text{PoS}(\mathcal{G}) \geq \frac{C(\text{PNE}(G_M))}{\text{OPT}(G_M)} > \frac{M}{2}.$$

By $M \to \infty$, this yields the desired result.

It remains to show $C(\text{PNE}(G_M)) > M$. For fixed $M \geq 1$, let $s = (z, p)$ be a PNE of G_M with induced Wardrop flow $x := x(s)$ and cost $K := K(s)$. Note that $z_i > 0$ holds for $i \in \{1, 2\}$, since any PNE has at least two positive capacities. Lemma 4.3.4

together with Theorem 4.3.6 yields that, for each firm $i \in \{1, 2\}$, the cost K either is optimal for $(P_i^1)(s_{-i})$, or optimal for $(P_i^2)(s_{-i})$, and has positive objective function value in both cases. Since for each firm $i \in \{1, 2\}$, the only candidate for a feasible solution of $(P_i^1)(s_{-i})$ is $2\sqrt{a_i \gamma_i} + b_i = \sqrt{a_i \gamma_i} + b_i + C_i$ and this yields an objective function value of 0, we get that K is an optimal solution of $(P_i^2)(s_{-i})$. In particular this yields, by considering firm 2, that $2M = \sqrt{a_2 \gamma_2} + b_2 + C_2 < K$. Furthermore we get $z_1 = \frac{x_1}{K-1}$ and $z_2 = \frac{M(1-x_1)}{K-M}$ for the capacities of the two firms. Altogether, the desired inequality for $C(s) = C(\mathrm{PNE}(G_M))$ follows:

$$C(\mathrm{PNE}(G_M)) = \frac{1}{z_1}x_1^2 + \frac{M}{z_2}(1 - x_1)^2 + z_1 + Mz_2 > (K - 1)x_1 + (K - M)(1 - x_1)$$
$$= Kx_1 - x_1 + K - Kx_1 - M + Mx_1 \geq K - M > M.$$

$\square$

4.7   Bibliographic Notes

The results presented in this chapter are a joint work with Tobias Harks, and an abstract appeared in the proceedings of the *15th International Conference on Web and Internet Economics (WINE 2019)* [64].

Conclusion

In this thesis, we analyzed problems exhibiting a two-stage structure, in which in the first stage, prices for using edges of a network are set, and in the second stage, selfish players use the edges and pay for it. The considered problems differ in particular in the way how the prices are set, namely by a central authority, or by selfishly acting firms.

First, we considered the problem of a system designer who chooses the cost sharing method for a second-stage network cost sharing game, and the system designer's objective is to minimize the PoS. The set of cost sharing methods from which the system designer is allowed to choose contains all budget-balanced, stable and separable cost sharing methods. Using the notion of an enforceable strategy profile (a profile which can actually appear as a PNE in a game induced by a choice of the system designer), the system designer's problem can be reformulated as finding an enforceable strategy profile with smallest possible cost. Note that given an enforceable strategy profile, our LP-characterization of enforceability also provides a cost sharing method enforcing the given profile. We mainly focused on conditions ensuring the existence of an optimal profile which is enforceable (or, in other words, conditions ensuring that an optimal solution of the system designer's problem yields a PoS of 1). For games with two players in undirected graphs, we completely characterized which graph structures ensure the existence of a globally optimal, enforceable solution, independent of the fixed edge costs (but with zero delays). A graph has this property if and only if there is no subgraph which is a Bad Configuration. An interesting open problem here is the computational complexity of recognizing whether a given graph contains a Bad Configuration as a subgraph. Further questions with respect to computational complexity, which we did not consider in this thesis, are if an optimal Steiner forest can be computed efficiently in a graph not containing a Bad Configuration, and if an optimal solution for the corresponding LP (and thus a cost sharing method enforcing it) from Section 3.2 can

A. Schedel, *Cost Sharing, Capacity Investment and Pricing in Networks*, Mathematische Optimierung und Wirtschaftsmathematik | Mathematical Optimization and Economathematics, https://doi.org/10.1007/978-3-658-33170-2_5

be computed in polynomial time. A characterization of efficient graphs for more than two players, or other more general settings, also remains an open problem. We already discussed why our proof for the two-player characterization cannot be used directly to derive a characterization for more general settings like directed graphs, more players, or other edge cost functions. Furthermore, our proof for the simple case with two players, an undirected graph, fixed edge costs and zero delays, is already quite long and complicated. Thus, a characterization like ours for more general settings is a challenging open problem for future research. For the general case with n players, we showed that if the given graph is n-series-parallel, i.e., the strategy set of each player is a series-parallel graph, then for nondecreasing, discrete concave edge costs and fixed, player-specific delays, every optimal solution is enforceable. We proved this by developing Algorithm n-SEPA, which takes as input an arbitrary strategy profile, and transforms this profile into an enforceable profile with cost at most the cost of the input profile. A corresponding cost sharing method enforcing the output profile is also provided by the algorithm. For the special case of constant edge costs, we were furthermore able to show that Algorithm n-SEPA can be executed in polynomial time. Therefore, Algorithm n-SEPA can be seen as a *black-box* reducing the efficient computation of a cost sharing method with low PoS to the efficient computation of a cheap solution for the underlying optimization problem. This way, a cost sharing method with PoS equal to 1 can be computed in polynomial time for the special case of an undirected n-series-parallel graph with fixed edge costs and zero delays (since for this setting, an optimal Steiner forest can be computed in polynomial time). It remains an open question whether Algorithm n-SEPA has polynomial running time also for the general case of nondecreasing, discrete concave edge costs. It would furthermore be interesting if besides the n-series-parallel graphs, other graph classes also allow for similar results. In general, to obtain a more realistic model, one should also allow for load-dependent delays in network cost sharing games.

Secondly, we studied a problem in which *selfishly* acting firms set prices (and capacities) for a second-stage network routing problem. Specifically, we considered networks consisting of parallel edges, and assumed that each firm controls exactly one edge. Furthermore, the price that a firm is allowed to choose is upper-bounded by a given firm-specific price cap. If there is at least one edge with positive capacity (after all firms have chosen capacities and prices), then subsequently one unit of flow is routed through the network, where each infinitesimally small customer wants to minimize the price plus the congestion cost of her chosen edge. The profit of a firm, which she wants to maximize, is derived from this routing problem and the capacity investment cost. We showed that for congestion cost functions which depend linearly on the volume of customers choosing a firm, and inverse-linearly on the edge capacity installed by the firm, there always exists a PNE in the described capacity and price competition game between the firms. Furthermore, the PNE is essentially unique, and compared to a social optimum, it's quality can get arbitrarily bad. The base for all these results is a characterization of the best response correspondences of the firms, which we also derived in this thesis. We completely characterized under which conditions a best response exists, and, in case of existence, showed that the optimal solutions of the corresponding bilevel optimization problem can be described by properties of two ordinary, 1-dimensional optimization problems. Due to the fact that best responses do not always exist, standard fixed-point arguments could not directly be applied to show existence of PNE. Instead, we used the concept of C-security introduced by McLennan et al. [89], combined with our characterization of best responses. An interesting direction for future research is whether results about existence, uniqueness and quality of PNE can be achieved for more general networks (e.g. series-parallel networks) and/or more general congestion cost functions. In particular, whether our approach via a characterization of the best responses of the firms combined with the result of McLennan at al. can be used to show existence of PNE. Regarding the quality of PNE, we showed that the *worst-case* PoA over all possible capacity and price competition games is unbounded. It would be interesting to further analyze the PoA of a specific instance, in particular, to derive upper bounds depending on the parameters of the instance. Finally, we think that a thorough analysis of the inelastic demand case is important for developing results for realistic models in which demand neither is strictly decreasing, nor constant. Think, for example, of a transportation model in which there is a preferred mode of transportation, but also a (less attractive) alternative. Then, it is a reasonable assumption that demand for the best mode stays constant for a certain price and congestion level until the alternative mode becomes attractive, and only then, demand for the best mode decreases. The analysis of such models is an interesting direction for future research.

Bibliography

1. D. Acemoglu, K. Bimpikis, and A. Ozdaglar. Price and capacity competition. *Games Econom. Behav.*, 66(1):1–26, 2009.
2. D. Acemoglu and A. Ozdaglar. Competition and efficiency in congested markets. *Math. Oper. Res.*, 32(1):1–31, 2007.
3. D. Acemoglu and A. Ozdaglar. Competition in parallel-serial networks. *IEEE Journal on Selected Areas in Communications*, 25(6):1180–1192, 2007.
4. H. Ackermann and A. Skopalik. On the complexity of pure Nash equilibria in player-specific network congestion games. In *Proc. 3rd Internat. Workshop on Internet and Network Economics*, pages 419–430, 2007.
5. C. D. Aliprantis and K. C. Border. *Infinite Dimensional Analysis*. Springer, Berlin Heidelberg, 1999.
6. E. Anshelevich and B. Caskurlu. Exact and approximate equilibria for optimal group network formation. *Theor. Comput. Sci.*, 412(39):5298–5314, 2011.
7. E. Anshelevich and B. Caskurlu. Price of stability in survivable network design. *Theory Comput. Syst.*, 49(1):98–138, 2011.
8. E. Anshelevich, A. Dasgupta, J. Kleinberg, É. Tardos, T. Wexler, and T. Roughgarden. The price of stability for network design with fair cost allocation. In *Proc. 45th Annual IEEE Sympos. Foundations Comput. Sci.*, pages 295–304, 2004.
9. E. Anshelevich, A. Dasgupta, J. Kleinberg, É. Tardos, T. Wexler, and T. Roughgarden. The price of stability for network design with fair cost allocation. *SIAM J. Comput.*, 38(4):1602–1623, 2008.
10. E. Anshelevich, A. Dasgupta, É. Tardos, and T. Wexler. Near-optimal network design with selfish agents. *Theory of Computing*, 4(1):77–109, 2008.
11. G. Avni and T. Tamir. Cost-sharing scheduling games on restricted unrelated machines. *Theor. Comput. Sci.*, 646:26–39, 2016.
12. M. Bateni, M. Hajiaghayi, and D. Marx. Approximation schemes for Steiner forest on planar graphs and graphs of bounded treewidth. *JACM*, 58(5):21:1–21:37, 2011.
13. M. Beckmann, C. McGuire, and C. Winsten. *Studies in the Economics and Transportation*. Yale University Press, New Haven, CT, USA, 1956.
14. C. Berge. *Topological Spaces: Including a Treatment of Multi-Valued Functions, Vector Spaces and Convexity*. Dover Publications, Mineola, New York, 1963.

© The Editor(s) (if applicable) and The Author(s), under exclusive license to
Springer Fachmedien Wiesbaden GmbH, part of Springer Nature 2021
A. Schedel, *Cost Sharing, Capacity Investment and Pricing in Networks*, Mathematische Optimierung und Wirtschaftsmathematik | Mathematical Optimization and Economathematics, https://doi.org/10.1007/978-3-658-33170-2

15. P. Bergendorff, D. Hearn, and M. Ramana. Congestion toll pricing of traffic networks. In P. Pardalos, D. Hearn, and W. Hager, editors, *Network Optimization*, pages 51–71, 1997.

16. P. Berman, A. Bhattacharyya, K. Makarychev, S. Raskhodnikova, and G. Yaroslavtsev. Approximation algorithms for spanner problems and directed Steiner forest. *Information and Computation*, 222:93–107, 2013.

17. V. Bilò and R. Bove. Bounds on the price of stability of undirected network design games with three players. *Journal of Interconnection Networks*, 12(1-2):1–17, 2011.

18. V. Bilò, I. Caragiannis, A. Fanelli, and G. Monaco. Improved lower bounds on the price of stability of undirected network design games. *Theory Comput. Syst.*, 52(4):668–686, 2013.

19. V. Bilò, M. Flammini, and L. Moscardelli. The price of stability for undirected broadcast network design with fair cost allocation is constant. In *Proc. 54th Annual IEEE Sympos. Foundations Comput. Sci.*, pages 638–647, 2013.

20. C. Bird. On cost allocation for a spanning tree: A game theoretic approach. *Networks*, 6:335–350, 1976.

21. K. C. Border. *Fixed Point Theorems with Applications to Economics and Game Theory*. Cambridge University Press, Cambridge, UK, 1985.

22. I. Caragiannis, V. Gkatzelis, and C. Vinci. Coordination mechanisms, cost-sharing, and approximation algorithms for scheduling. In *Proc. 13th Internat. Conf. Web and Internet Economics*, pages 74–87, 2017.

23. J. Cardinal and M. Hoefer. Non-cooperative facility location and covering games. *Theor. Comput. Sci.*, 411:1855–1876, March 2010.

24. H.-L. Chen and T. Roughgarden. Network design with weighted players. *Theory Comput. Syst.*, 45(2):302–324, 2009.

25. H.-L. Chen, T. Roughgarden, and G. Valiant. Designing network protocols for good equilibria. *SIAM J. Comput.*, 39(5):1799–1832, 2010.

26. X. Chen, Z. Diao, and X. Hu. Network characterizations for excluding Braess's paradox. *Theory Comput. Syst.*, 59(4):747–780, 2016.

27. G. Christodoulou, C. Chung, K. Ligett, E. Pyrga, and R. van Stee. On the price of stability for undirected network design. In *7th International Workshop on Approximation and Online Algorithms*, pages 86–97, 2009.

28. G. Christodoulou, V. Gkatzelis, and A. Sgouritsa. Cost-sharing methods for scheduling games under uncertainty. In *Proc. 18th ACM Conf. Economics and Computation*, pages 441–458, 2017.

29. G. Christodoulou, S. Leonardi, and A. Sgouritsa. Designing cost-sharing methods for bayesian games. In *Proc. 9th Intl. Symp. Algorithmic Game Theory*, pages 327–339, 2016.

30. G. Christodoulou and A. Sgouritsa. Designing networks with good equilibria under uncertainty. In *Proc. of the 27th Annual ACM-SIAM Symposium on Discrete Algorithms*, pages 72–89, 2016.

31. R. Cominetti, J. Correa, and N. Stier-Moses. The impact of oligopolistic competition in networks. *Oper. Res.*, 57(6):1421–1437, 2009.

32. J. Correa, C. Guzmán, T. Lianeas, E. Nikolova, and M. Schröder. Network pricing: How to induce optimal flows under strategic link operators. In *Proc. of the 2018 ACM Conference on Economics and Computation*, pages 375–392, 2018.

33. J. Correa and N. Stier-Moses. *Wardrop Equilibria*. Wiley Encyclopedia of Operations Research and Management Science, 2011.
34. X. Deng, T. Ibaraki, and H. Nagamochi. Algorithmic aspects of the core of combinatorial optimization games. *Math. Oper. Res.*, 24(3):751–766, 1999.
35. R. Dial. Minimal-revenue congestion pricing part I: A fast algorithm for the single-origin case. *Transportation Res.*, 33(B):189–202, 1999.
36. R. Dial. Network-optimized road pricing: Part I: A parable and a model. *Oper. Res.*, 47(1):54–64, 1999.
37. Y. Disser, A. E. Feldmann, M. Klimm, and M. Mihalák. Improving the H_k-bound on the price of stability in undirected shapley network design games. *Theor. Comput. Sci.*, 562:557–564, 2015.
38. M. Englert, T. Franke, and L. Olbrich. Sensitivity of Wardrop equilibria. In *Proc. 1st Internat. Sympos. Algorithmic Game Theory*, pages 158–169, 2008.
39. A. Fabrikant, C. Papadimitriou, and K. Talwar. The complexity of pure Nash equilibria. In *Proc. 36th Annual ACM Sympos. Theory Comput.*, pages 604–612, 2004.
40. K. Fan. Fixed point and minmax theorems in locally convex topological linear spaces. *Proc. Natl. Acad. Sci. USA*, 38:121–126, 1952.
41. A. Fanelli, D. Leniowski, G. Monaco, and P. Sankowski. The ring design game with fair cost allocation. *Theoretical Computer Science*, 562:90–100, 2015.
42. U. Feige. A threshold of ln n for approximating set cover. *J. ACM*, 45(4):634–652, 1998.
43. M. Feldman, G. Kortsarz, and Z. Nutov. Improved approximation algorithms for directed Steiner forest. *J. Comput. Syst. Sci.*, 78:279–292, 2012.
44. M. Feldman and T. Tamir. Conflicting congestion effects in resource allocation games. *Oper. Res.*, 60(3):529–540, 2012.
45. L. Fleischer, K. Jain, and M. Mahdian. Tolls for heterogeneous selfish users in multi-commodity networks and generalized congestion games. In *Proc. 45th Annual IEEE Sympos. Foundations Comput. Sci.*, pages 277–285, 2004.
46. S. Fortune, J. Hopcroft, and J. Wyllie. The directed subgraph homeomorphism problem. *Theoret. Comput. Sci.*, 10:111–121, 1980.
47. R. Freeman, S. Haney, and D. Panigrahi. On the price of stability of undirected multicast games. In *Proc. 12th Internat. Conf. on Web and Internet Economics*, pages 354–368, 2016.
48. M. Gairing, T. Harks, and M. Klimm. Complexity and approximation of the continuous network design problem. *SIAM Journal on Optimization*, 27(3):1554–1582, 2017.
49. M. Gairing, K. Kollias, and G. Kotsialou. Tight bounds for cost-sharing in weighted congestion games. In *Proc. 42nd Internat. Coll. on Automata, Languages, and Programming*, pages 626–637, 2015.
50. M. Gairing, K. Kollias, and G. Kotsialou. Cost-sharing in generalised selfish routing. In *Proc. 10th Internat. Conference on Algorithms and Complexity*, pages 272–284, 2017.
51. V. Gkatzelis, K. Kollias, and T. Roughgarden. Optimal cost-sharing in general resource selection games. *Oper. Res.*, 64(6):1230–1238, 2016.
52. I. Glicksberg. A further generalization of the Kakutani fixed point theorem, with application to Nash equilibrium points. *Proc. Amer. Math. Soc.*, 3:170–174, 1952.
53. M. X. Goemans and M. Skutella. Cooperative facility location games. *J. Algorithms*, 50(2):194–214, 2004.

54. R. Gopalakrishnan, J. R. Marden, and A. Wierman. Potential games are *Necessary* to ensure pure Nash equilibria in cost sharing games. *Math. Oper. Res.*, 39(4):1252–1296, 2014.

55. D. Granot and G. Huberman. On minimum cost spanning tree games. *Math. Prog.*, 21:1–18, 1981.

56. D. Granot and M. Maschler. Spanning network games. *Internat. J. Game Theory*, 27:467–500, 1998.

57. S. Guha and S. Khuller. Greedy strikes back: Improved facility location algorithms. *J. Algorithms*, 31:228–248, 1999.

58. T. D. Hansen and O. Telelis. On pure and (approximate) strong equilibria of facility location games. In *Proc. 4th Internat. Workshop on Internet and Network Economics*, pages 490–497, 2008.

59. T. D. Hansen and O. Telelis. Improved bounds for facility location games with fair cost allocation. In *Proc. 3rd Intl. Conf. Combinatorial Optimization and Applications*, pages 174–185, 2009.

60. T. Harks, M. Hoefer, A. Huber, and M. Surek. Efficient black-box reductions for separable cost sharing. In *Proc. 45th Internat. Coll. on Automata, Languages, and Programming*, pages 154:1–154:15, 2018.

61. T. Harks, M. Hoefer, A. Schedel, and M. Surek. Efficient black-box reductions for separable cost sharing. *Math. Oper. Res.*, 2020 (to appear).

62. T. Harks, A. Huber, and M. Surek. A characterization of undirected graphs admitting optimal cost shares. In *Proc. 13th Internat. Conf. Web and Internet Economics*, pages 237–251, 2017.

63. T. Harks and B. Peis. Resource buying games. *Algorithmica*, 70(3):493–512, 2014.

64. T. Harks and A. Schedel. Capacity and price competition in markets with congestion effects. In *Proc. 15th Internat. Conference on Web and Internet Econom.*, page 341, 2019.

65. T. Harks, A. Schedel, and M. Surek. A characterization of undirected graphs admitting optimal cost shares. *SIAM J. Disc. Math.*, 33(4):1932–1996, 2019.

66. T. Harks, M. Schröder, and D. Vermeulen. Toll caps in privatized road networks. *European Journal of Operational Research*, 276(3):947–956, 2019.

67. T. Harks and P. von Falkenhausen. Optimal cost sharing for capacitated facility location games. *European Journal of Operational Research*, 239(1):187–198, 2014.

68. D. Hochbaum. Heuristics for the fixed cost median problem. *Math. Prog.*, 22:148–162, 1982.

69. M. Hoefer. Non-cooperative tree creation. *Algorithmica*, 53:104–131, 2009.

70. M. Hoefer. Competitive cost sharing with economies of scale. *Algorithmica*, 60:743–765, 2011.

71. M. Hoefer. Strategic cooperation in cost sharing games. *Internat. J. Game Theory*, 42(1):29–53, 2013.

72. M. Hoefer and P. Krysta. Geometric network design with selfish agents. In *Proc. 11th Internat. Conf. Computing and Combinatorics*, pages 167–178, 2005.

73. F. K. Hwang, D. S. Richards, and P. Winter. The Steiner tree problem. *Annals of Discrete Mathematics*, volume 53, 1992.

74. K. Jain and M. Mahdian. Cost sharing. In N. Nisan, T. Roughgarden, E. Tardos, and V. Vazirani, editors, *Algorithmic Game Theory*, chapter 15, pages 385–410. Cambridge University Press, Cambridge, UK, 2007.

75. R. Johari, G. Y. Weintraub, and B. Van Roy. Investment and market structure in industries with congestion. *Oper. Res.*, 58(5):1303–1317, 2010.

76. S. Kakutani. A generalization of Brouwer's fixed point theorem. *Duke Mathematics Journal*, 8(3):457–458, 1941.

77. L. Khachiyan. Polynomial algorithms in linear programming. *USSR Computational Mathematics and Mathematical Physics*, 20(1):53–72, 1980.

78. M. Klimm and D. Schmand. Sharing non-anonymous costs of multiple resources optimally. In *Proc. 9th International Conference on Algorithms and Complexity*, pages 274–287, 2015.

79. K. Kollias and T. Roughgarden. Restoring pure equilibria to weighted congestion games. *ACM Trans. Economics and Comput.*, 3(4):1–21, 2015.

80. B. Korte and J. Vygen. *Combinatorial Optimization: Theory and Algorithms*. Springer, Berlin, Germany, 2012.

81. E. Koutsoupias and C. Papadimitriou. Worst-case equilibria. In *Proc. 16th Internat. Sympos. Theoretical Aspects of Comput. Sci.*, pages 404–413, 1999.

82. J. Li. An $o(\log n \log \log n)$ upper bound on the price of stability for undirected shapley network design games. *Information Processing Letters*, 109(15):876–878, 2009.

83. S. Li. A 1.488 approximation algorithm for the uncapacitated facility location problem. *Inf. Comput.*, 222:45–58, 2013.

84. T. Litman. Understanding transport demands and elasticities: How prices and other factors affect travel behavior. http://www.vtpi.org/elasticities.pdf, 2019. Accessed: 2019-22-10.

85. T.-L. Liu, J. Chen, and H.-J. Huang. Existence and efficiency of oligopoly equilibrium under toll and capacity competition. *Transportation Research Part E: Logistics and Transportation Review*, 47(6):908–919, 2011.

86. A. Mamageishvili, M. Mihalák, and S. Montemezzani. Improved bounds on equilibria solutions in the network design game. *International Journal of Game Theory*, 47(4):1113–1135, 2018.

87. P. Marcotte. Network design problem with congestion effects: A case of bilevel programming. *Math. Program., Ser. A*, 34:142–162, 1986.

88. Y. Marinakis, A. Migdalas, and P. M. Pardalos. Cost allocation in combinatorial optimization games. In A. Chinchuluun, P. M. Pardalos, A. Migdalas, and L. Pitsoulis, editors, *Pareto Optimality, Game Theory And Equilibria*, pages 217–247. Springer New York, New York, 2008.

89. A. McLennan, P. Monteiro, and R. Tourky. Games with discontinuous payoffs: A strengthening of Reny's existence theorem. *Econometrica*, 79(5):1643–1664, 2011.

90. N. Megiddo. Cost allocation for Steiner trees. *Networks*, 8(1):1–6, 1978.

91. D. Monderer and L. Shapley. Potential games. *Games Econom. Behav.*, 14(1):124–143, 1996.

92. K. Nishimura and J. Friedman. Existence of Nash equilibrium in n person games without quasi-concavity. *International Economic Review*, 22(3):637–648, 1981.

93. M. Osborne and A. Rubinstein. *A Course in Game Theory*. MIT Press, Cambridge, MA, USA, 1994.

94. C. Papadimitriou. Algorithms, games, and the Internet. In *Proc. 33th Annual ACM Sympos. Theory Comput.*, pages 749–753, 2001.
95. P. Reny. On the existence of pure and mixed strategy Nash equilibria in discontinuous games. *Econometrica*, 67(5):1029–1056, 1999.
96. J. Rich, O. Kveiborg, and C. Hansen. On structural inelasticity of modal substitution in freight transport. *Journal of Transport Geography*, 19(1):134–146, 2011.
97. R. Rosenthal. A class of games possessing pure-strategy Nash equilibria. *Internat. J. Game Theory*, 2:65–67, 1973.
98. T. Roughgarden and O. Schrijvers. Network cost-sharing without anonymity. *ACM Trans. Economics and Comput.*, 4(2):8, 2016.
99. T. Roughgarden and É. Tardos. How bad is selfish routing? *J. ACM*, 49(2):236–259, 2002.
100. D. Schmand, A. Skopalik, and M. Schröder. Network investment games with Wardrop followers. *arXiv*, arXiv:1904.10417, 2019.
101. V. Syrgkanis. The complexity of equilibria in cost sharing games. In *Proc. 6th Internat. Workshop on Internet and Network Econom.*, pages 366–377, 2010.
102. A. Tamir. On the core of network synthesis games. *Math. Prog.*, 50:123–135, 1991.
103. P. von Falkenhausen and T. Harks. Optimal cost sharing for resource selection games. *Math. Oper. Res.*, 38(1):184–208, 2013.
104. J. Wardrop. Some theoretical aspects of road traffic research. *Proc. Inst. Civil Engineers*, 1(Part II):325–378, 1952.
105. F. Xiao, H. Yang, and D. Han. Competition and efficiency of private toll roads. *Transportation Research Part B: Methodological*, 41(3):292–308, 2007.
106. H. Yang and H.-J. Huang. The multi-class, multi-criteria traffic network equilibrium and systems optimum problem. *Transportation Research Part B: Methodological*, 38(1):1–15, 2004.
107. H. Yang, F. Xiao, and H. Huang. Private road competition and equilibrium with traffic equilibrium constraints. *Journal of Advanced Transportation*, 43(1):21–45, 2009.

Printed in the United States
by Baker & Taylor Publisher Services

Printed in the United States
by Baker & Taylor Publisher Services